Monographs in Electrical and Electronic Engineering

Editors: P. HAMMOND, D. WALSH

Monographs in Electrical and Electronic Engineering

R. L. Bell: *Negative electron affinity devices* (1973)
A. A. Bergh and P. J. Dean: *Light-emitting diodes* (1975)
G. Gibbons: *Avalanche-diode microwave oscillators* (1973)
G, S. Hobson: *The Gunn effect* (1974)
R. L. Stoll: *The analysis of eddy currents* (1974)
H. C. Wright: *Infrared techniques* (1973)

Noise and fluctuations

in electronic devices and circuits

F. N. H. ROBINSON

CLARENDON PRESS · OXFORD
1974

Oxford University Press, Ely House, London W. 1

GLASGOW NEW YORK TORONTO MELBOURNE WELLINGTON
CAPE TOWN IBADAN NAIROBI DAR ES SALAAM LUSAKA ADDIS ABABA
DELHI BOMBAY CALCUTTA MADRAS KARACHI LAHORE DACCA
KUALA LUMPUR SINGAPORE HONG KONG TOKYO

ISBN 0 19 859319 8

PRINTED IN NORTHERN IRELAND
AT THE UNIVERSITIES PRESS
BELFAST

Preface

PHYSICISTS and engineers are interested in noise in electronic systems for essentially the same reason. It is easy to build an electronic system whose sensitivity is limited not by incidental imperfections in its construction but solely by the more fundamental limitations imposed by the atomic structure and statistical behaviour of matter. To the physicist noise in an electronic system represents an important practical manifestation of the phenomena described by statistical mechanics, and an understanding of its practical consequences helps to illuminate and clarify the meaning of some of the concepts and results of this abstract branch of physical theory. To the engineer noise is an obstacle to the practical realization of useful devices, and an understanding of its physical origins helps him to minimize its effects by informed and careful design. With these two groups of readers in mind I have attempted to write a book which, because it relates the practical effects of noise to basic physical laws, will be acceptable and interesting to the physicist and yet bc sufficiently practical and comprehensive to be accessible and useful to the engineer.

A complete description of noise in electronic systems requires not only an account of the origins of noise but also of the statistical properties of noise and their influence on the detection of threshold signals and the optimal encoding and processing of signals. Many books on noise deal almost exclusively with the second group of topics, but here these topics are given scant attention. The emphasis is entirely on the physical origins of noise in electronic devices and the way noise behaves in those simple electronic circuits which form the building blocks of complete systems.

In writing this book I have had the benefit of many stimulating discussions with Dr. G. A. Brooker of the Clarendon Laboratory, and I am greatly indebted to Mr. D. F. Gibbs of Bristol University, whose careful reading of the text and whose pertinent and illuminating comments have helped me to eliminate many errors and statements of doubtful validity or impenetrable obscurity.

F. N. H. ROBINSON

Clarendon Laboratory
Oxford
August, 1973

Contents

Introduction

In everyday speech noise, in Ambrose Bierce's phrase, is 'a stench in the ear', but to the electronic engineer it also signifies any random background against which coherent signal information has to be distinguished. The complexity and refinement, and especially the degree of amplification, attainable with electronic apparatus are such that the background generally has its origin in the random statistical behaviour of the atomic constituents of matter. The ultimate sensitivity is then set by the fundamental laws which govern the behaviour of matter and radiation. Only this aspect of noise will concern us, and we shall ignore all those miscellaneous hissing, frying, and crackling noises which arise from interference or faulty components.

This book is intended primarily for the electronic engineer or experimental scientist who has to design and operate sensitive electronic equipment and who wishes to understand the limitations imposed on its performance by the existence of noise. The bulk of the book is concerned with the origins and behaviour of noise in the principal components and sub-assemblies of electronic systems. It is not concerned with what may be termed the grand strategy of electronic system design, i.e. problems such as those concerned with the optimal encoding or processing of signals. The material to be discussed provides the necessary background information required for making such strategic decisions.

Noise has a, not entirely undeserved, reputation for difficulty and obscurity. This is due partly to the complexity of systems in which noise is significant and partly to the unfamiliar nature of some of the mathematical tools involved. There is, however, another reason—this is the way the study of noise calls on the results of three rather disparate intellectual disciplines, statistical mechanics, circuit theory, and statistical analysis. The fundamental processes which give rise to noise in electronic components arise from phenomena which form the subject matter of statistical mechanics, but the way noise enters and propagates through an electronic system requires the techniques of circuit theory for its analysis. Finally, when, as a result of this analysis, we have reached a description of the behaviour of the output of a system we have to interpret this description in statistical terms.

For most of the time we shall be using the language of circuit theory, where the emphasis is on the frequency or spectral components of the signal and noise in a system, and for this purpose the most useful concept is that of a noise power spectrum. If $I(t)$ represents a time-dependent current containing signals and noise we describe the signal components of $I(t)$ as

$$I_s(t) = \int_0^\infty I_s(f)\exp(2\pi jft)\, df,$$

and the noise is described by the departure $\Delta I(t) = I(t) - I_s(t)$ of the current from this expected or average behaviour. The noise power spectrum $w(f)$ is defined so that the contribution to $\{(\Delta I(t))^2\}$ from spectral components in df at f is $w(f)\, df$. In this language a complete statement about the output of a noisy system would be contained in expressions for $I_s(f)$ and $w(f)$. In the course of the calculation of the output we would begin with the signal spectrum of the input, then identify the noise power spectra associated with the individual components of the system, and go on to discover how individual signal and noise components are propagated through the system. Finally, we should combine these components, taking due account of any correlation, or lack of correlation, to give the output spectra. In statistical language we would then consider $I_s(t)$ as giving the expected output at t and

$$\langle(\Delta I(t))^2\rangle = \int w(f)\, df$$

as giving the mean square deviation of the output from this expected value.

Clearly, this procedure implies that we have some fairly precise notion of what we mean by the expected, mean, or average values of quantity, and this is something that we shall discuss in Chapter 1. Equally, this notion that we are dealing with statistical statements is connected with the random nature of the fundamental noise processes and with the relation between noise as an experimentally observable process in time, and the fluctuations discussed when matter and radiation is described in terms of statistical and quantum mechanics. Although, in all cases of practical interest, it is possible, using simple statistical or thermodynamic arguments, to obtain the basic results about noise required for circuit analysis without invoking any more fundamental considerations, we deal with this problem in some detail in Chapters 5 and 6. However, because these chapters may not be easily accessible, or of interest, to the reader whose intentions are severely practical and because conventional treatments of statistical mechanics do not usually deal with these topics, we now make some general remarks about the connection between noise and statistical mechanics.

A complete specification of the microscopic state of an object such as a 47 Ω resistor would involve giving the coordinates and momenta of all the 10^{20} or so atomic and subatomic particles which are involved in its construction. Not only does quantum mechanics restrain us from trying to do this but, even if this restraint is neglected, the sheer magnitude of the task puts it beyond the bounds of practical possibility. Instead, we have to be content with a much cruder description of the resistor in terms of its chemical composition and mechanical construction, together with the fact that it is in equilibrium at atmospheric pressure and temperature. Nevertheless, statistical mechanics still allows us to make some predictions about its behaviour. In particular we can calculate the probability with which any possible microscopic configuration of the particles will occur. Since each microscopic configuration corresponds to a particular value of the macroscopically observable quantities associated with the resistor, this allows us to calculate the probability that one of the quantities, say the short-circuit current through the resistor, has a value between I and $I+\mathrm{d}I$. This probability is the sum of the probabilities of all those microscopic configurations which lead to a current in this range. We can then go on to calculate the expected value of the current $\langle I\rangle$ and the expected mean square deviations $\langle\Delta I^2\rangle$ about this value. Now these results have the following interpretation, $\langle I\rangle$ gives the expected or average short-circuit current to be observed if we make a series of measurements on a large set (or ensemble) of identical resistors and $\langle\Delta I^2\rangle$ gives the mean square deviation of these results in this ensemble. It is, perhaps, also reasonable that, if we have but a single resistor, $\langle I\rangle$ and $\langle\Delta I^2\rangle$ might also refer to the results of a set of measurements of I taken at widely separated instants on this one resistor. However, by themselves the results do not tell us anything about how I will vary with time over a continuous interval. To obtain this information we have to introduce other considerations. Although we cannot specify the exact configuration of the particles at any instant, these particles are subject to quite precise dynamical laws, and these laws govern the evolution of one microscopic configuration into another, as time proceeds. Thus given that at t_1 the system is in one of the group of microscopic configurations that lead to a current I_1 we can at least make statistical predictions about the probability that, at a later time t_2, it will be found in a group of configurations that lead to a current I_2. In other words we can describe the evolution of I in time in statistical terms. Armed with this information we can then go on to make statistical estimates of the behaviour of the spectral or Fourier components of I and so derive the power spectrum. This process is spelt out, in some detail, in Chapter 5, and as we shall see it leads to a rather clear distinction between those devices, e.g. the field-effect transistor, in which noise is appropriately treated as a thermal process, and others, e.g. the bipolar transistor, in which it is a shot noise process. In Chapter 5 the currents and voltages are treated in classical terms although the particles of the system are regarded as basically subject to quantal laws. Chapter 6 considers the consequences of treating the currents and voltages as subject to quantal laws.

With the exception of these chapters and Chapter 3, where the basic mathematical techniques required later are developed, little demand is made on the reader's mathematical ability. Even in Chapter 3 a glib acquiescence to Fourier series and integrals, rather than a profound understanding of them, is all that is required.

Chapters 1–7 deal with general topics and the fundamental sources of noise, and Chapters 8–11 with noise in the main active devices used at low to moderately high frequencies. Chapters 12–14 then deal with noise in amplifiers constructed from these devices, while Chapters 15–17 treat u.h.f. and microwave amplifying devices. The next four chapters discuss non-linear circuits and devices, and finally, in the last chapter, some applications chosen partly for their intrinsic interest and partly for purposes of illustration are discussed. This chapter concludes with a brief note on the practical design of low-noise equipment.

The list of references at the end of the book is followed by a brief general bibliography.

1 Signals and Noise

1.1. Fluctuations

IF in attempting to measure a small steady voltage, say a thermal e.m.f., we use a d.c. amplifier and meter we find that up to a point we can detect smaller and smaller signals by increasing the gain of the amplifier, but that eventually the sensitivity is limited by random fluctuations of the meter. A single reading then is insufficient to tell us the magnitude of the voltage since part of the reading may well have been a fluctuation. However, we could still estimate the voltage if we first performed a series of experiments to establish the zero reading and the magnitude of the fluctuations.

Thus suppose that with no input we obtain a series of readings $N_i(i = 1 \ldots n)$ while with a known input we obtain $S_j(v)(j = 1 \ldots n)$. We can then form the averages

$$\langle N \rangle = \frac{1}{n} \sum_{i=1}^{n} N_i \tag{1.1a}$$

and

$$\langle S(v) \rangle = \frac{1}{n} \sum_{j=1}^{n} S_j(v), \tag{1.1b}$$

and the mean square deviations

$$\langle \Delta N^2 \rangle = \frac{1}{n} \sum (N_i - \langle N \rangle)^2 \tag{1.2a}$$

$$\langle \Delta S^2(v) \rangle = \frac{1}{n} \sum (S_j(v) - \langle S(v) \rangle)^2, \tag{1.2b}$$

and it is reasonable to suppose that as n becomes large each of these averages tends to a definite limit, so that

$$\langle V(v) \rangle = \langle S(v) \rangle - \langle N \rangle \tag{1.3}$$

represents the effect of the input v. From these readings we could also estimate the probability that a single reading $S_j(v)$ departs from $\langle S(v) \rangle$ by more than, say, $\langle \Delta S^2(v) \rangle^{\frac{1}{2}}$.

It is generally assumed in physics that fluctuating quantities are normally distributed about their mean value, and in many cases there are sound theoretical arguments, based on the central limit theorem, for this assumption; but in any case, in almost every instance it is exceedingly unlikely that any single reading will depart from $\langle S(v)\rangle$ by more than a few times $\langle \Delta S^2(v)\rangle^{\frac{1}{2}}$. We therefore adopt a root mean square (r.m.s.) criterion of probable error, saying that a single reading $S_j(v)$ gives $\langle S(v)\rangle \pm \langle \Delta S^2(v)\rangle^{\frac{1}{2}}$. If we take m readings the sum of the squares of the errors will increase as m while the square of the sum of the readings will increase as m^2, so that the fractional error in an average over m readings decreases, i.e.

$$\langle S(v)\rangle_m = \langle S(v)\rangle \pm \left(\frac{\langle \Delta S^2(v)\rangle}{m}\right)^{\frac{1}{2}}. \tag{1.4}$$

If, after calibrating the instrument, we use it to measure an unknown voltage u and find in a single reading $S = \langle S(v)\rangle$, we can then say that

$$u = v \pm \delta v,$$

where

$$\frac{\delta v}{v} = \frac{\langle \Delta S^2(v)\rangle^{\frac{1}{2}}}{\langle S(v)\rangle - \langle N\rangle}. \tag{1.5}$$

The inverse of this ratio we refer to as the signal to noise ratio S/N. The fractional accuracy of a single reading of the meter is N/S. The purpose of noise theory is to provide a basis for calculating results such as (1.5) and for discussing the dependence of the signal to noise ratio on the parameters of the system.

It should be especially noted that we define S/N as an amplitude and not a power ratio. The converse practice, though more usual, leads to confusion, especially when small changes in large signals must be considered.

1.2. Ensembles

Noise is a random statistical process and statements about noise are probability statements. These are most easily handled using the concept of a statistical ensemble, drawn from statistical mechanics. If we were told that a certain length of wire had a resistance of 100 Ω and was maintained at a pressure of 1 atm and a temperature T, we would recognize this as a reasonably complete macroscopic description of a wire-wound 100 Ω resistor. However, we would realize also that the description was insufficient to specify the exact configuration of all the atoms and electrons in the wire, and that there would be many microscopically distinct configurations of these particles which were consistent with the given macroscopic description. In some of these configurations the voltage, or potential difference, across the ends of the resistor might be non-zero and for one group of possible configurations might exceed some value θ, whereas for all other possible configurations it would be less than θ. We could also describe this by saying that if we had an ensemble of identical resistors specified in an identical way then for some fraction $\pi(\theta)$ of all the resistors in the ensemble the voltage would be less than θ and for a fraction $1-\pi(\theta)$ it would be greater than θ. Now in statistical mechanics and in its application to noise calculations, we assume that a proper statistical ensemble contains one copy of the original system in every single possible microscopic configuration compatible with the over-all macroscopic description. Thus $\pi(\theta)$ is equal to the fraction of all possible microscopic configurations that yield a voltage less than θ. The average voltage observed across the ends of all the resistors in the ensemble is the ensemble average

$$\langle\theta\rangle = \int \theta \, \mathrm{d}\pi(\theta), \tag{1.6a}$$

and this is also the expectation value of the voltage for a resistor chosen at random from the ensemble. In other words it is the expected voltage across the ends of the one actual resistor available to us. The mean square fluctuations in the voltage, i.e. the measure of the possible spread in the values of the voltage about the mean or expected value, are given by

$$\langle\Delta\theta^2\rangle = \int (\theta-\langle\theta\rangle)^2 \, \mathrm{d}\pi(\theta), \tag{1.6b}$$

and so, for our one available resistor, we expect the voltage to be around $\langle\theta\rangle$ and not to differ from this value by more than a few times $\langle\Delta\theta^2\rangle^{\frac{1}{2}}$. Implicit in the introduction of these concepts is the notion that, once the physical constitution of the system is known, the probability distribution function $\pi(\theta)$, or at least the values of $\langle\theta\rangle$ and $\langle\Delta\theta^2\rangle^{\frac{1}{2}}$, should be calculable.

We can generalize this idea to apply to more complex systems. If the physical construction of an electronic instrument and the set of all external parameters (e.g. supply voltage, temperature, and input voltage) which specify its state are known, then these parameters define a definite statistical ensemble to which the instrument belongs. This ensemble consists of identical copies of the instrument subject to identical macroscopic external conditions. Any external property of the instrument, e.g. its output voltage, will depend, however, not only on all these given parameters but also on the exact microscopic internal state of the instrument, and each copy of the instrument in the ensemble will lead to a different output. If a fraction $\pi(\theta)$ of the members of the ensemble leads to an output recognizable as less than θ then the expected output of the instrument is the ensemble average (1.6a) and the fluctuations in the output are given by (1.6b). By relating probability statements about noise to ensemble averages we make contact with the theoretical results of statistical mechanics and thermodynamics and, as a result, we can often derive the results we require by simple physical arguments. Thus, for example, as we shall see in Chapter 4, we can show that the expectation value of the voltage across the ends of a resistance R in thermal equilibrium at a temperature T is zero but, if the voltage is observed using an otherwise noiseless instrument of bandwidth $\delta\nu$, the mean square fluctuations about this mean are $4RkT\delta\nu$, where $k = 1{\cdot}38\times10^{-23}$ JK^{-1} is Boltzmann's constant. In other words if we make measurements of the voltage at different instants of time the average of the measurements will be zero but the average of the squares of the measurements will be $4RkT\delta\nu$. In practical electronics we should interpret this as meaning that the resistance exhibited a fluctuating e.m.f. with this mean square value.

1.3. Averages

Theoretically derived ensemble averages apply to many systems observed at one time. A group of observations on one system made at many times is equivalent to the results for an ensemble only if the times are equivalent, i.e. specified in the same way with respect to any known parameters of the system and its input; and if they are so far apart that all 'memory' of the state of the system is obliterated between measurements. We have tacitly assumed this to be the case in § 1.1. We shall consider later when this assumption is justified.

In any case, ensemble averages, which can be calculated theoretically, must be distinguished from time averages, which apply to a continuous observation of one system. We shall always denote ensemble averages by $\langle\ \rangle$ and time averages by bars. Thus

$$\bar{I} = \lim_{T\to\infty} \frac{1}{T}\int_0^T I(t)\,\mathrm{d}t. \tag{1.7a}$$

We shall also need averages over a finite period t to $t+\tau$ which we denote

$$\overline{I^\tau} = \frac{1}{\tau}\int_t^{t+\tau} I(t)\,\mathrm{d}t. \tag{1.7b}$$

In many cases a single observation of the output of a system of finite bandwidth or resolving time furnishes an average such as $\overline{I^\tau}$ and we shall then be interested in mixed averages, e.g.

$$\langle\overline{I^\tau}\rangle = \frac{1}{n}\sum_{j=1}^{n}\frac{1}{\tau}\int_t^{t+\tau} I_j(t)\,\mathrm{d}t, \tag{1.8}$$

where $I_j(t)$ is the current in the jth system of the ensemble at t.

The essential difference between ensemble and time averages is most clearly seen by considering a coherent signal $I_0 \cos \omega t$ for which

$$\langle I(t)\rangle = I_0 \cos \omega t$$

and $\bar{I} = 0$ while

$$\overline{I^\tau} = (\sin \omega(t+\tau) - \sin \omega t)I_0/\omega\tau.$$

1.4. Stationary processes

A stationary process obeys statistical laws and is subject to conditions that do not vary with time. All times are then equivalent and a group of observations at times $t_1 \ldots t_j \ldots t_n$ forms an ensemble provided that for all i and j, $|t_i - t_j| > \tau$, where τ is a time longer than any characteristic time of the process or the system by which it is observed. Thus if $\theta(t)$ represents the effect of a stationary process, we can write the time average $\bar{\theta}$ in the form

$$\bar{\theta} = \frac{1}{T}\int_0^T \left\{\frac{1}{N}\sum_{k=1}^{N} \theta_k(t_k + t')\right\} dt',$$

where the times t_k are spaced τ apart. Each term in the bracket is an ensemble. average of θ and, since the process is stationary, does not depend on the time; thus

$$\bar{\theta} = \frac{1}{T}\int_0^T \langle\theta\rangle \, dt' = \langle\theta\rangle, \tag{1.9}$$

and so for a stationary process time and ensemble averages are equivalent.

The three fundamental sources of noise in electronic apparatus, i.e. Johnson, shot, and flicker noise, are stationary processes.

1.5. Observations

No observation of a physical quantity is instantaneous; either we explicitly and consciously average, say, a meter reading over a finite time or we use a system with a finite response time. In most cases we do both, and in addition a subjective element creeps in owing to such factors as the persistence of vision, etc. In order to interpret the results of noise theory we need a definition of a single observation which eliminates so far as possible all undetermined psychological and physiological factors. We therefore define a single observation as one made in a time short compared with the natural time constants of the system, i.e. short compared with the time required for the output to change significantly. A visual reading of a meter whose time constant is several seconds is a single observation, as is a photograph of a cathode ray tube (c.r.t.) trace taken in a time short compared with the time base period (it will show only a single spot of light). On the other hand, a visual observation of the screen, probably occupying several complete traces, or a reading of a short period meter are not single observations.

Noise calculations yield the average value and the statistical fluctuations of single readings defined in this way. In some cases, especially when the indicating instrument is a meter, the response time of the entire system is that of the meter and is exactly calculable, in other cases, e.g. with c.r.t. or audio indicators, the response time is partly determined by the physical system preceding the indicator, partly by the indicator, and partly by the observer. To avoid having to consider the ill-defined properties of the indicator and the observer, we therefore adopt the convention that the signal and noise to be calculated are those in the final output of the system immediately prior to the indicator, i.e. at the meter terminals, the y plates, or the loudspeaker coil. In this way we separate the properties of the physical instrument from those of the observer.

1.6. Periodic signals

So far we have considered only systems whose expected output is constant, and we have been able to maintain a consistently statistical approach. When the signal input varies with time the expected output signal will do so also, and the signal to noise ratio will no longer be constant. This may be regarded either as a fundamental difficulty or as merely an inconvenience. From the latter standpoint, if $S(t)$ is the signal output we might adopt an r.m.s. definition of the signal and define the signal to noise ratio as

$$\frac{S}{N} = \left(\frac{\overline{S^2(t)}}{\langle \Delta V^2 \rangle}\right)^{\frac{1}{2}}, \tag{1.10}$$

where $\langle \Delta V^2 \rangle$ are the statistical fluctuation of the output as previously defined Eq. (1.10) then gives the expected signal to noise ratio in a single measurement but averaged over all possible phases of the signal. From this point of view we regard the output as a continuous process whose value at any instant is of interest and, in principle, measureable. The difficulty arises when we collate a series of measurements and attempt by Fourier analysis to disentangle the different frequency components of the output and assign amplitudes and probable errors to them. We can only do this if the measurements are numerous and extend over a considerable time, and we then no longer have to deal with a single instantaneous measurement. This problem is considered in more detail in § 3.4.

In one sense, however, this problem is irrelevant since in almost every case of practical interest we observe static output signals. The one main exception is an audible signal, which we shall consider separately. Thus if the output of a system consists of a signal $S(t)$ with components at f_0 and its harmonics mf_0, we almost invariably present it on an oscilloscope with a time base of frequency f_0. The resulting output is a static trace, and, if the form of the signal is known, only one number, its amplitude, is needed to characterize it. This number can be obtained in a true single observation, made in a short time compared with the response time of the system to changes in the signal amplitude, and so the statistical considerations appropriate to d.c. signals apply to it. The details of the calculation of the expected value of the output and its fluctuations are, of course, different (see § 3.5).

Eqn (1.10), apart from the square root, coincides with the definition used in most writings on the subject. We have chosen to make our fundamental choice of S/N an amplitude rather than the more usual power ratio since the latter choice does not correspond to any simple criterion of observability and leads to confusion in considering the detectability of small changes in large signals.

Audible output signals present a different problem. For complex signals such as music the ear appears to respond to the mean signal power and to regard the noise as a background; eqn (1.10) then provides a direct criterion of signal to noise ratio. On the other hand, the ear can recognize very weak sounds of a single frequency against a considerable background of noise, and effectively acts as a filter (see § 3.4). The problem is in any case more one of physiology than physics and we shall not discuss it further.

1.7. The structure of noise calculations

The main purpose of calculating the noise in an electronic system is to discover what is the least signal it will detect or what is the accuracy with which a large signal can be measured, and to show how the result depends on the parameters of the system.

We shall show in succeeding chapters that if a system accepts signals in a bandwidth Δf from a resistive source at a temperature T the least signal power that we can detect under the most favourable circumstances is

$$P_{\min} = kT\,\Delta f, \tag{1.11a}$$

where k is Boltzmann's constant, and that if the source resistance is R the least change δV in the open circuit voltage is

$$\delta V = (4RkT\,\Delta f)^{\frac{1}{2}}. \tag{1.11b}$$

It follows that the least change δP in a large power P is

$$\delta P = 2(PkT\,\Delta f)^{\frac{1}{2}}. \tag{1.11c}$$

With any actual system this sensitivity is not achieved and so we introduce a figure of merit for the system F, the noise figure, so that

$$P_{\min} = FkT\,\Delta f, \tag{1.12a}$$

$$\delta V = (4RFkT\,\Delta f)^{\frac{1}{2}}, \tag{1.12b}$$

$$\delta P = 2(PFkT\,\Delta f)^{\frac{1}{2}}. \tag{1.12c}$$

In any but the simplest systems Δf is not identical with the signal bandwidth of the component parts of the system. Thus, as we shall show in Chapter 10, if we have an r.f. amplifier of bandwidth Δf_1 followed by a detector and an audio amplifier of bandwidth Δf_2 much less than Δf_1, then the appropriate value of Δf is eqn (1.11a) is

$$\Delta f = (2\,\Delta f_1\,\Delta f_2)^{\frac{1}{2}}. \tag{1.13}$$

From the signal point of view Δf_2 corresponds to the maximum permissible rate of change of the signal amplitude, while Δf_1 corresponds to the permissible range of carrier frequencies.

In eqns (1.11) and (1.12) we have assumed that a signal is detectable when the change in the output due to the signal is equal to noise, and so these equations are related to considerations about signal to noise ratio. We see that we can specify the sensitivity of any system to signals of a specified form and strength by giving just two parameters for the system F and Δf, and so noise calculations should be designed to yield these two parameters. The reason why we do not amalgamate them in one parameter $F\Delta f$ is that in many cases F and Δf have simple physical meanings. Thus in nearly every case F is associated with noise introduced in the early stages of the system and is a measure of the quality of the system, while Δf is associated with the structure of the system. In the illustration above Δf was determined by the information that the system consisted of amplifiers of bandwidth Δf_1, and $\Delta f_2 \leqslant \Delta f_1$ separated by a detector. F for this system would depend on the detailed design of the input stage of the first amplifier.

2 Temperature-Limited Shot Noise

THE fluctuations in the current in a temperature-limited diode form one of the simplest examples of a noise current and therefore, despite their relative practical unimportance, we shall use them as an introduction to the techniques of noise theory.

Whenever a circuit contains a potential barrier, such as that at a rectifying junction in a transistor or at the cathode surface in a vacuum tube, the current is limited to those electrons that have sufficient thermal energy to surmount the potential barrier. The current therefore fluctuates in a way determined by the normal thermal fluctuations in the position and energy distribution of the electrons. In many cases the elementary features of this type of noise, shot noise, are obscured by space charge and transit time, but in a close spaced temperature limited diode with a high enough anode voltage it is possible to neglect these effects. Each electron emitted by the cathode is then immediately drawn to the anode and contributes to the current in the external circuit.

2.1. The Poisson distribution

If we take a sufficiently short interval of time δt, the probability that an electron is emitted in δt is small and proportional to δt:

$$p = \nu\,\delta t. \tag{2.11}$$

The probability that two or more electrons are emitted in δt is negligible. Thus if we consider a finite time τ and divide it into very many equal intervals δt such that $\tau = N\,\delta t$, the chance that exactly m electrons are emitted in m specified intervals δt is $p^m(1-p)^{N-m}$. The intervals can be chosen in ${}^N\mathrm{C}_m$ ways and so the probability that m electrons are emitted in τ is

$$P(m,\tau) = {}^N\mathrm{C}_m p^m q^{N-m}, \tag{2.2}$$

where we have written $q = 1-p$.

In an ensemble of intervals the average number emitted will be

$$\langle m\rangle = \sum_{m=0}^{N} mP(m,\tau), \tag{2.3}$$

and the mean square number

$$\langle m^2\rangle = \sum m^2 P(m,\tau). \tag{2.4}$$

These sums are readily evaluated, for example,

$$\langle m\rangle = \sum {}^{N}C_m m p^m q^{N-m} = p\frac{\partial}{\partial p}(p+q)^N$$

$$= Np(p+q)^{N-1} = Np, \tag{2.5}$$

and

$$\langle m^2\rangle = N^2p^2+Np-Np^2. \tag{2.6}$$

As we take the limit $N\to\infty$ then $p\to 0$ but Np remains finite. The last term in eqn (2.6) may be neglected therefore and we have

$$\langle\Delta m^2\rangle \equiv \langle m^2\rangle-\langle m\rangle^2 = Np = \langle m\rangle. \tag{2.7}$$

This is the well-known result for the fluctuations in a Poisson distribution.

2.2. Application to shot noise

In Fig. 2.1 we show a circuit with which it is, in principle, possible to verify eqn (2.7) and obtain a value for the charge on the electron. During a time τ

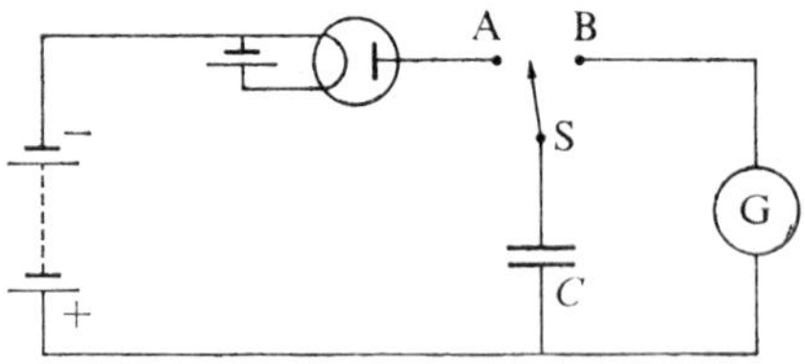

FIG. 2.1. Measurement of shot noise with a ballistic galvanometer.

the switch S is connected to A and the current through the diode charges C. At the end of this interval S is connected to B and the charge measured ballistically. If e is the charge on an electron, the measured charge is

$$Q = em,$$

so that, in a series of measurements,

$$\langle Q\rangle = e\langle m\rangle,$$

$$\langle\Delta Q^2\rangle = e^2\langle\Delta m^2\rangle = e^2\langle m\rangle = e\langle Q\rangle.$$

We might also regard each measurement as yielding the average current flowing during τ,

$$\overline{I^\tau} = Q/\tau = em/\tau.$$

If we do this we obtain

$$\langle(\overline{\Delta I^\tau})^2\rangle = \frac{e}{\tau}\langle\overline{I^\tau}\rangle = \frac{e}{\tau}I. \tag{2.8}$$

Thus the fluctuations in each measurement of I decrease as τ increases.

2.3. Campbell's theorem

We now consider a slightly more sophisticated arrangement, shown in Fig. 2.2, in which the current is continuously recorded by a critically damped galvanometer of period τ and deflection $\theta = kI$ for a current I. We shall require a theorem due to Campbell (1909) which states that if an instrument responds to an indefinitely short pulse of charge q at $t = 0$ by a deflection

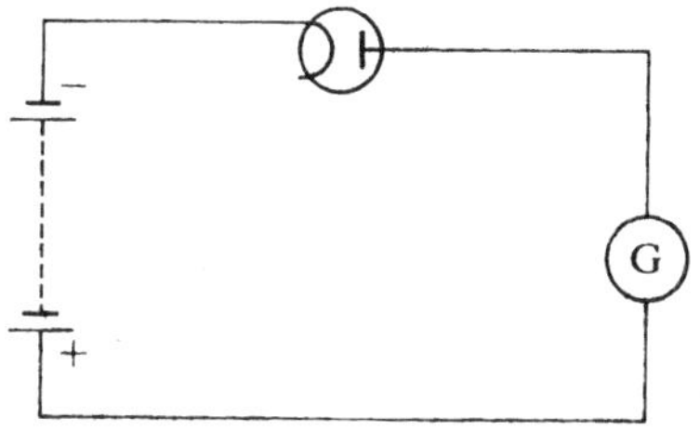

FIG. 2.2. Measurement of shot noise with a critically damped galvanometer.

$\theta(t) = qF(t)$ then its mean response to a random series of pulses occurring at a mean rate ν is

$$\langle\theta\rangle = \nu q \int_{-\infty}^{\infty} F(t)\,\mathrm{d}t \tag{2.9a}$$

with fluctuations

$$\langle\Delta\theta^2\rangle = \nu q^2 \int_{-\infty}^{\infty} F^2(t)\,\mathrm{d}t. \tag{2.9b}$$

The response of a critically damped galvanometer to a pulse q is well known to be

$$\theta(t) = qk\left(\frac{2\pi}{\tau}\right)^2 t\exp\left(-\frac{2\pi t}{\tau}\right) \quad \text{for} \quad t > 0.$$

The pulses in a shot noise current have a magnitude $q = e$ and occur at a mean rate $\nu = I_0/e$, where I_0 is the direct current. Thus Campbell's theorem yields

$$\langle\theta\rangle = I_0\int_0^{\infty} k\left(\frac{2\pi}{\tau}\right)^2 t\exp\left(-\frac{2\pi t}{\tau}\right)\mathrm{d}t = kI_0$$

and

$$\langle\Delta\theta^2\rangle = eI_0\int_0^{\infty} k^2\left(\frac{2\pi}{\tau}\right)^4 t^2\exp\left(-\frac{4\pi t}{\tau}\right)\mathrm{d}t = \frac{\pi k^2 eI_0}{2\tau}. \tag{2.10}$$

The first result need occasion no comment but the second result shows that the fluctuations, expressed as current fluctuations, have a mean square value

$$\langle\Delta I^2\rangle = \frac{\pi eI_0}{2\tau}. \tag{2.11}$$

If we compare this with eqn (2.8) we see that a single instantaneous reading of the galvanometer is equivalent to an average of the current over the preceding interval of length $2\tau/\pi$.

To prove Campbell's theorem we consider an ensemble of systems over a long interval of length T. During this long interval a single measurement of θ is made on each system at some definite time t, the same for each system in the ensemble. From the ensemble we then select a sub-ensemble consisting of those systems in which exactly m pulses have arrived during the interval T. In one interval, from this sub-ensemble, let the times of arrival of the pulses be $t_1 \ldots t_k \ldots t_m$, then the response at t is

$$\theta_m(t) = q\sum_{k=1}^{m} F(t-t_k). \tag{2.12a}$$

The arrival times t_k are randomly distributed over T so that the probability that t_k lies in $\mathrm{d}t'$ at t' is just $\mathrm{d}t'/T$. Thus the sub-ensemble average of $\theta_m(t)$ is

$$\langle\theta(t)\rangle_m = q\int_0^T \frac{\mathrm{d}t_1}{T}\ldots\int_0^T \frac{\mathrm{d}t_m}{T}\sum_{k=1}^{m} F(t-t_k).$$

This gives

$$\langle\theta(t)\rangle_m = q\sum_{k=1}^{m}\int_0^T F(t-t_k)\frac{\mathrm{d}t_k}{T} = qm\int_0^T F(t-t')\frac{\mathrm{d}t'}{T}$$

$$= qm\int_{t-T}^{t} F(t')\frac{\mathrm{d}t'}{T}. \tag{2.12b}$$

Now let T be large compared with Δ, the range of times for which $F(t)$, the effect of a single pulse, persists. Then for all except a fraction $2\Delta/T$ of the times t in T we shall have $\Delta < t < T-\Delta$ and so we can, without significant error, replace the limits in eqn (2.12b) by $\pm\infty$:

$$\langle\theta(t)\rangle_m = \frac{qm}{T}\int_{-\infty}^{\infty} F(t')\,\mathrm{d}t'. \tag{2.13a}$$

We now average over all values of m in the ensemble, note that $\nu = \langle m\rangle/\tau$, and obtain the first part of Campbell's theorem (eqn (2.9a)).

The second part of the theorem is proved by evaluating

$$\theta_m^2(t) = q^2\sum_{k=1}^{m}\sum_{l=1}^{m} F(t-t_k)F(t-t_l).$$

The sub-ensemble average is

$$\langle\theta^2(t)\rangle_m = q^2\sum\sum\int\frac{\mathrm{d}t_1}{T}\ldots\int\frac{\mathrm{d}t_m}{T}F(t-t_k)F(t-t_l).$$

In this sum there are m terms with $k = l$ and $m(m-1)$ with $k \neq l$, and so

$$\langle\theta^2(t)\rangle_m = \frac{q^2m}{T}\int_0^T F^2(t-t')\,\mathrm{d}t' + \frac{q^2m(m-1)}{T^2}\left\{\int_0^T F(t-t')\,\mathrm{d}t\right\}^2. \tag{2.13b}$$

Dealing with the limits as before, we have

$$\langle\theta^2(t)\rangle_m = \frac{q^2m}{T}\int_{-\infty}^{\infty} F^2(t)\,\mathrm{d}t + \frac{q^2m(m-1)}{T^2}\left\{\int_{-\infty}^{\infty} F(t)\,\mathrm{d}t\right\}^2. \tag{2.14}$$

We now take the ensemble average over m and use our earlier result (eqn (2.7)) $\langle m^2\rangle = \langle m\rangle + \langle m\rangle^2$ and obtain

$$\langle\theta^2\rangle = q^2\nu\int_{-\infty}^{\infty} F^2(t)\,\mathrm{d}t + \langle\theta\rangle^2. \tag{2.15}$$

With the identity $\langle\Delta\theta^2\rangle = \langle\theta^2\rangle - \langle\theta\rangle^2$ this immediately yields the second part of Campbell's theorem (eqn (2.9b)).

Campbell's theorem yields the fluctuations in an ensemble of instantaneous readings of any instrument responding to shot noise if the instrument can be specified by a response function of the form $\theta(t) = qF(t)$. It leads, as we have seen, to a relatively simple treatment of galvanometers.

2.4. Frequency response

The majority of electronic systems have, however, a response which is most easily measured and economically expressed as a frequency response curve. Thus if the input is a harmonic current $J(f)\exp(2\pi \mathrm{j} ft)$ the output is

$$\phi(f)\exp(2\pi \mathrm{j} ft) = G(f)J(f)\exp(2\pi \mathrm{j} ft).$$

The response function $F(t)$ is well known to be simply related to $G(f)$. Thus let

$$I(t) = \int_{-\infty}^{\infty} J(f)\exp(2\pi \mathrm{j} ft)\,\mathrm{d}t$$

be an arbitrary input current, then we may express the output as either

$$\theta(t) = \int_{-\infty}^{\infty} F(t-t_1)I(t_1)\,\mathrm{d}t_1$$

or as

$$\theta(t) = \int_{-\infty}^{\infty} G(f)J(f)\exp(2\pi \mathrm{j} ft)\,\mathrm{d}f.$$

If we now express $J(f)$ as

$$J(f) = \int_{-\infty}^{\infty} I(t_1)\exp(-2\pi \mathrm{j} f t_1)\, \mathrm{d}t_1,$$

we obtain

$$\int_{-\infty}^{\infty} F(t-t_1)I(t_1)\, \mathrm{d}t_1 = \int_{-\infty}^{\infty} I(t_1) \int_{-\infty}^{\infty} G(f)\exp(2\pi \mathrm{j} f(t-t_1))\, \mathrm{d}f\, \mathrm{d}t_1$$

and, since $I(t_1)$ is arbitrary, this leads to

$$F(t) = \int_{-\infty}^{\infty} G(f)\exp(2\pi \mathrm{j} f t)\, \mathrm{d}f. \tag{2.16}$$

Thus $F(t)$ is the Fourier transform of $G(f)$ and we can use Parseval's theorem, i.e.

$$\int_{-\infty}^{\infty} F^2(t)\, \mathrm{d}t = \int_{-\infty}^{\infty} G(f)G^*(f)\, \mathrm{d}f.$$

Now $F(t)$ is real and so $G^*(f) = G(-f)$ and $G(f)G^*(f)$ is an even function of f. Therefore we can avoid discussing negative frequencies and write

$$\int_{-\infty}^{\infty} F^2(t)\, \mathrm{d}t = 2\int_{0}^{\infty} G(f)G^*(f)\, \mathrm{d}f \tag{2.17}$$

This step is equivalent to the convention in linear theory that we represent $\cos \omega t$ by $\mathrm{Re}(\exp(\mathrm{j}\omega t))$ and not by

$$\tfrac{1}{2}(\exp(\mathrm{j}\omega t)+\exp(-\mathrm{j}\omega t)).$$

We can write Campbell's theorem as

$$\langle \Delta\theta^2 \rangle = 2\nu q^2 \int_{-\infty}^{\infty} G(f)G^*(f)\, \mathrm{d}f, \tag{2.18}$$

in which form it can be used to discuss shot noise in conventional electronic circuits containing inductances and capacitances.

Thus, if a temperature-limited current of mean value I_0 flows in the input of an amplifier whose current gain is $G(f)$, the output fluctuations are

$$\Delta\theta^2 = 2eI_0 \int_{0}^{\infty} GG^*\, \mathrm{d}f. \tag{2.19}$$

2.5. Shot noise in a tuned amplifier

To illustrate this result we consider the arrangement of Fig. 2.3. If a voltage input V to the amplifier results in a c.r.t. deflection $\theta = KV$ cm, then the mean reading of θ will be zero and the gain of the system at frequency f is

$$\frac{\theta(f)}{J(f)} = G(f) = \frac{KR}{1+jQ(f/f_0-f_0/f)}, \tag{2.20}$$

where $Q = 2\pi f_0 CR$ and $(2\pi f_0)^2 LC = 1$. We therefore have

$$\int_0^\infty GG^* \, df = \frac{\pi K^2 R^2 f_0}{2Q},$$

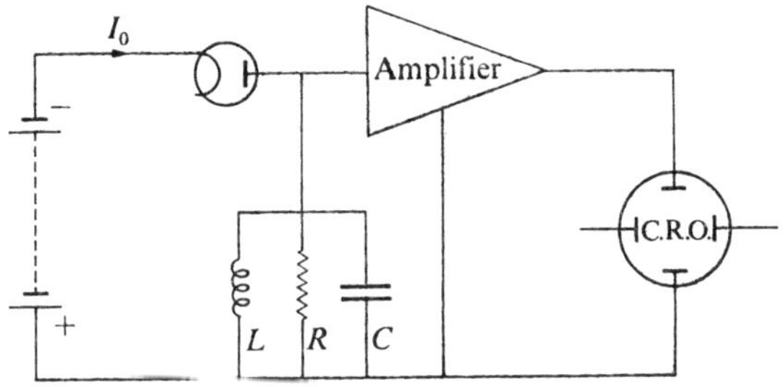

FIG. 2.3. Shot noise input to a tuned circuit.

and so individual readings of θ will fluctuate about zero with a mean square deviation

$$\Delta\theta^2 = \pi e I_0 R^2 K^2 \frac{f_0}{Q}. \tag{2.21}$$

If, instead, a signal current $J(f)$ is impressed on the circuit the response is given by eqn (2.20). The (time) mean square value of $J(f)$ is $\frac{1}{2}J(f)J^*(f)$, and the signal to noise ratio at the centre frequency f_0 is

$$\frac{S}{N} = \left(\frac{JJ^*Q}{2\pi e I_0 f_0}\right)^{\frac{1}{2}}. \tag{2.22}$$

The signal to noise ratio for an equal signal at other frequencies will be less, of course.

The form of eqn (2.19) suggests that it would be possible to regard the shot noise current as exhibiting a spectrum of fluctuations of which the mean square value associated with the frequency range df is

$$dI_n^2 = 2eI_0 \, df. \tag{2.23}$$

We would then obtain the correct result if we assumed that fluctuations of different frequencies were uncorrelated.

If we adopt this point of view we can also write eqn (2.22) in the form

$$\frac{S}{N}=\left(\frac{\frac{1}{2}JJ^*}{2eI_0\,\Delta f_n}\right)^{\frac{1}{2}}, \tag{2.24}$$

where the effective noise bandwidth of the system is defined as

$$\Delta f_n=\frac{\int\limits_0^\infty GG^*\,\mathrm{d}f}{G(f_0)G^*(f_0)}=\frac{\pi}{2}\frac{f_0}{Q}. \tag{2.25}$$

We can further express this result by saying that the noise in the system is equivalent to a fluctuating current of mean square value $2eI_0\,\Delta f_n$ at the input.

This representation of the effect of noise in terms of a spectral density $2eI_0$ of the noise source and an effective bandwidth of the system is very common, because many systems have a well-defined bandwidth in which $G(f)=G_0$ within Δf and is virtually zero elsewhere. In these cases the effective noise bandwidth of the system coincides with the signal bandwidth.

In the circuit of Fig. 2.3 the useful signal bandwidth is usually defined as that half-power points. The signal and noise bandwidths are then related by

$$\Delta f_s=\frac{2}{\pi}\Delta f_n. \tag{2.26}$$

We may also remark that for a signal near the edge of the signal bandwidth the output signal to noise ratio will be $1/\sqrt{2}$ times that given by eqn (2.24).

2.6. Square-law detection of shot noise

Let us now consider what happens when the oscilloscope in Fig. 2.3 is replaced by a square-law meter whose response to a voltage V applied to the amplifier is $\theta=\alpha V^2$. The average response of the meter is then $\langle\theta\rangle=\alpha\langle V^2\rangle$, and we have

$$\langle\Delta V^2\rangle=2eI_0R^2\,\Delta f_n.$$

Thus

$$\langle\theta\rangle=\alpha.2eI_0R^2\,\Delta f_n,$$

so that the mean meter reading is proportional to the mean square fluctuations in Δf_n.

If the meter time constant is τ, each reading of the meter is an average over the preceding τ seconds. In τ seconds we can make $\tau\,\Delta f$ independent readings of the output of a system of time constant $1/\Delta f$. Thus each meter reading corresponds to an average of $\tau\,\Delta f$ measurements of the noise. The fluctuations in the meter reading are therefore reduced in mean square value by this ratio:

$$\langle \Delta\theta^2 \rangle \sim \frac{\langle\theta\rangle^2}{\tau\,\Delta f}.$$

We can thus obtain $\langle\theta\rangle$ to a fractional accuracy $(\tau\,\Delta f)^{-\frac{1}{2}}$ and so, if we wish to measure shot noise fluctuations, we should choose a long time constant meter preceded by a wide-band amplifier.

In Chapter 20 we shall return to this problem and give a more detailed treatment.

2.7. Gaussian noise

The mean square value of the noise is, for most purposes, a useful measure of its intensity, but occasionally we also need to know the statistical distribution of the noise, i.e. the probability or fraction of the time that the noise amplitude exceeds some particular value. Thus, for example, if we know that the expected r.m.s. noise output of an amplifier is $\langle\Delta V^2\rangle^{\frac{1}{2}}$ we would also wish to know how much bigger than $\langle\Delta V^2\rangle^{\frac{1}{2}}$ we have to make the dynamic range of the amplifier to ensure that it is not overloaded by peaks in the noise. It is possible to show (Rice 1944) that the two most common sources of noise in electronic systems, i.e. shot noise and thermal noise (see Chapter 4), both have a Gaussian distribution. Thus in any linear system in which noise is generated by these processes the probability $\mathrm{d}P$, or fraction of the time, that a noise voltage of mean square value $\langle\Delta V^2\rangle$ is within the range V to $V+\mathrm{d}V$ is

$$\mathrm{d}P = \left(\frac{1}{2\pi\langle\Delta V^2\rangle}\right)^{\frac{1}{2}}\exp\left(-\frac{V^2}{2\langle\Delta V^2\rangle}\right)\mathrm{d}V. \tag{2.27}$$

The result, of course, does not hold for non-linear systems. For example, if the noise output of a linear system is passed through a square-law device the resulting output will have the noise peaks accentuated and the smaller fluctuations attenuated.

We shall not attempt to give the rather difficult proof of (2.27) here, but rather concentrate on its consequences. The probability that the magnitude of the noise voltage exceeds V is

$$P_{>V} = \int_{-\infty}^{-V}\frac{\mathrm{d}P}{\mathrm{d}V}\,\mathrm{d}V + \int_{V}^{\infty}\frac{\mathrm{d}P}{\mathrm{d}V}\,\mathrm{d}V = 2\int_{V}^{\infty}\frac{\mathrm{d}P}{\mathrm{d}V}\,\mathrm{d}V, \tag{2.28}$$

and, if we express V in terms of its ratio to the r.m.s. noise as

$$V = x\langle\Delta V^2\rangle^{\frac{1}{2}}, \tag{2.29}$$

this yields

$$P_{>V} = \left(\frac{2}{\pi}\right)^{\frac{1}{2}}\int_{x}^{\infty}\exp(-u^2/2)\,\mathrm{d}u = \mathrm{erfc}(x/\sqrt{2}), \tag{2.30}$$

where erfc is the complementary error function. Some useful values of $P_{>V}$ are as follows:

$x = 0$	0·68	1	2	4	10
$P = 1$	0·5	0·32	0·045	6×10^{-5}	10^{-23}.

We see that although the magnitude exceeds $\langle\Delta V^2\rangle^{\frac{1}{2}}$ for about a third of the time it very rarely exceeds $4\langle\Delta V^2\rangle^{\frac{1}{2}}$ and virtually never (for much less than a pico-second per century) exceeds $10\langle\Delta V^2\rangle^{\frac{1}{2}}$. If therefore an amplifier is designed to handle about 10 times the expected r.m.s. noise there is virtually no chance of its being overloaded by peaks in the noise.

The very rapid decrease of $P_{>V}$ with increasing V means that once a signal appreciably exceeds the r.m.s. noise, it can be detected with almost absolute certainty. Thus if the output of an amplifier suddenly rises to 4 times the r.m.s. noise level there is less than one chance in 10^4 that this increase is due to noise alone.

3 Power Spectra

IN the last chapter we saw that the fluctuations in the output of a system of gain $G(f)$ responding to a shot-noise input could be expressed as

$$\langle\Delta\theta^2\rangle = \int_0^\infty G(f)G^*(f)w(f)\,\mathrm{d}f, \tag{3.1}$$

where $w(f) = 2eI_0$. This corresponds to a continuous spectral distribution of fluctuations in the shot-noise input. Since, in this case, $w(f)$ is equal to the power dissipated in unit resistance by the Fourier components of the fluctuating current in unit frequency interval about f, it is known as the power spectrum; and this terminology is used even when the noise process is not a current. We consider now how a general fluctuating quantity can be described in these terms.

3.1. Definition

Let $I(t)$ be any time-dependent quantity, coherent or random, defined or observed, over a period T, then it can be expressed as a Fourier series of fundamental period T, e.g. as

$$I(t) = \sum_{n=-\infty}^{\infty} x_n \exp\left(\mathrm{j}\frac{2\pi nt}{T}\right), \tag{3.2}$$

where, because $I(t)$ is real, $x_n^* = x_{-n}$. If, in addtion, $I(t)$ is known to represent a stationary, random process the expectation values of $I(t)$, $I^2(t)$, etc. must all be independent of t. Now

$$\langle I(t)\rangle = \sum_{n=-\infty}^{\infty} \langle x_n\rangle \exp\left(\mathrm{j}\frac{2\pi nt}{T}\right) \tag{3.3}$$

and, if this is to be independent of t, the expectation values of the Fourier coefficients x_n except for $\langle x_0\rangle$ must vanish. Thus

$$\langle I(t)\rangle = \langle x_0\rangle. \tag{3.4}$$

We also have

$$\langle I^2(t)\rangle = \sum_m \sum_n \langle x_m x_n^*\rangle \exp\left\{\mathrm{j}\frac{2\pi(m-n)t}{T}\right\}, \tag{3.5}$$

and, if this is to be independent of t, all the expectation values $\langle x_m x_n^* \rangle$ must vanish except when $m = n$, so that

$$\langle I^2(t) \rangle = \sum_{n=-\infty}^{\infty} \langle x_n x_n^* \rangle = \langle x_0^2 \rangle + 2\sum_{n=1}^{\infty} \langle x_n x_n^* \rangle. \tag{3.6}$$

The coefficient x_0 is the time-average value of $I(t)$ over the period T and then, if we take T large enough, in any reasonable situation we can assume that x_0 has the same value for all systems in the statistical ensemble used to form the expectation values. Thus $\langle x_0^2 \rangle = \langle x_0 \rangle^2$ and the expected mean square fluctuations are

$$\langle \Delta I^2 \rangle = \langle I^2 \rangle - \langle I \rangle^2 = 2\sum_{n=1}^{\infty} \langle x_n x_n^* \rangle. \tag{3.7}$$

As T is made indefinitely long, the interval $1/T$ between the frequencies of successive components becomes infinitesimal and so, provided that $\langle x_n x_n^* \rangle$ varies smoothly with n, we can define a spectral density

$$w(f) = \lim_{T\to\infty} 2T\langle x_n x_n^* \rangle, \tag{3.8}$$

where $f = n/T$, and write (3.7) as

$$\langle \Delta I^2 \rangle = \int_0^{\infty} w(f)\, \mathrm{d}f. \tag{3.9}$$

If $I(t)$ is presented as the input to a system of gain $G(f) = G(n/T)$, the output will be

$$\theta(t) = \sum_{n=-\infty}^{\infty} G\left(\frac{n}{T}\right) x_n \exp\left(\mathrm{j}\frac{2\pi nt}{T}\right), \tag{3.10}$$

and, since G is not a random variable, we obtain

$$\langle \theta(t) \rangle = G(0)\langle x_0 \rangle$$

and

$$\langle \Delta\theta^2 \rangle = 2\sum_{n=1}^{\infty} G\left(\frac{n}{T}\right) G^*\left(\frac{n}{T}\right) \langle x_n x_n^* \rangle,$$

which can be expressed as

$$\langle \Delta\theta^2 \rangle = \int_0^{\infty} G(f)G^*(f)w(f)\, \mathrm{d}f,$$

with $w(f)$ defined by (3.8). Thus, for a stationary, random process this is an adequate definition of the power spectrum.

If $I(t)$ contains, in addition to the random component, a known coherent or signal component $S(t)$, we can obviously apply the same arguments to the random component $I(t)-S(t)$, but it is sometimes convenient to regard $w(f)$ defined by (3.8) as applying to any type of time-dependent quantity. This is especially the case in a long, formal, mathematical discussion of signals and noise in a complex system. It does lead however to difficulties. If, for example, $I(t)$ contains a coherent term

$$A \cos 2\pi f_0 t = \tfrac{1}{2} A \exp(2\pi \mathrm{j} f_0 t) + \tfrac{1}{2} A \exp(-2\pi \mathrm{j} f_0 t)$$

then $\langle x_n x_n^* \rangle$ is no longer a smoothly varying function of n and changes discontinuously at $n = f_0 T$. The passage from the sum (3.7) to the integral (3.9) is then no longer permissible, and leads to a delta-function singularity $\frac{1}{2}A^2\delta(f-f_0)$ in $w(f)$. The area under this singularity is $\frac{1}{2}A^2$ and equal to the time-average signal power. An alternative approach is to relate the sum (3.7) not to a Riemann integral such as (3.9) but to a Stieltjes integral, so that $\int_0^f \mathrm{d}P(f)$ is the power associated with Fourier components up to a frequency f. The cumulative power spectrum $P(f)$ remains finite but has discontinuities at the frequencies of coherent signals. The response of a system of gain $G(f)$ is then given by

$$\langle \Delta\theta^2 \rangle = \int_{f=0}^{\infty} G(f)G^*(f)\,\mathrm{d}P(f), \qquad (311)$$

but the contributions to $\langle \Delta\theta^2 \rangle$ from discontinuities in $P(f)$ are no longer fluctuation terms and correspond to signals. The mean output signal power is the sum of these discontinuous contributions, and the output noise arises only from the continuous part of P.

So far we have assumed implicitly that $I(t)$ corresponds to a low-level noise process at the input of an electronic system, but clearly we can also consider the power spectrum $W(F)$ of the output $\theta(t)$, and then the mean square fluctuations in the output are

$$\langle \Delta\theta^2 \rangle = \int_0^{\infty} W(F)\,\mathrm{d}F. \qquad (3.12)$$

Thus, if $W(F)$ is known, we can calculate the random spread in the values that will be obtained for the instantaneous output. Normally this would represent the final output of a calculation of the over-all noise performance of an electronic system. Thus a typical noise calculation consists of identifying and calculating the power spectra $w_i(f)$ of the various noise processes due to system components, and then calculating the contribution each such process makes to the output-noise power spectrum. The remaining chapters of this book are devoted mainly to these two problems.

In some cases, e.g. thermal or Johnson noise in resistors, physical arguments lead directly to an expression for the noise power spectrum to be associated with the component, but in other cases, shot noise discussed in

Chapter 2 is a typical example, the initial calculation yields not $w(f)$ but the temporal statistics of the noise. We then need a mathematical procedure for obtaining $w(f)$. The relevant theorems are usually known as inversion formulae and are discussed in the next section.

Although, for most purposes, the complex exponential Fourier-series representation of $I(t)$ given by (3.2) is adequate, when we come to deal with non-linear systems it leads to difficulties, and it is desirable therefore to replace it by

$$I(t) = c_0 + \sum_{n=1}^{\infty} c_n \cos\left(\frac{2\pi nt}{T} + \phi_n\right). \tag{3.13}$$

It is then easy to see, since $x_n = \frac{1}{2}c_n \exp(j\phi_n)$, that the definition of $w(f)$ becomes

$$w(f) = \lim_{T\to\infty} \frac{T}{2}\langle c_n^2\rangle. \tag{3.14}$$

Also occasionally it is convenient to use both sine and cosine terms so that

$$I(t) = a_0 \sum_{n=1}^{\infty} \left\{a_n \cos\left(\frac{2\pi nt}{T}\right) + b_n \sin\left(\frac{2\pi nt}{T}\right)\right\}. \tag{3.15}$$

In this case

$$w(f) = \lim_{T\to\infty} \frac{T}{2}\langle a_n^2 + b_n^2\rangle \tag{3.16}$$

and, for a stationary random process,

$$\langle a_n^2\rangle = \langle b_n^2\rangle. \tag{3.17}$$

3.2. Inversion formulae

Before discussing these formulae it will be useful to indicate, by means of an example, the type of calculation in which they are used. If two contacts are made to a specimen of a semiconductor there will be a fluctuating current $I(t)$ in the external circuit due to charge carriers being generated and recombining in the material. A carrier generated at a specific time g_k, which pursues a specific path p_k and recombines at a specific time r_k, will make a calculable contribution $i_k(t)$ to $I(t)$, and $I(t)$ will consist of a sum of such contributions:

$$I(t) = \sum_k i_k(t).$$

Any function of $I(t)$ can be expressed in terms of these individual contributions. In particular we can form either

$$I(t)I(t+\tau) = \sum_k \sum_l i_k(t) i_l(t+\tau)$$

or

$$I^2(t) = \sum_k \sum_l i_k(t) i_l(t)$$

and, if we wish, take the time averages of these quantities over either indefinitely long periods of time or over finite periods of time. Statistical mechanics will also furnish us with the statistical properties of the generation times g_k, recombination times r_k, and paths p_k. These determine the individual values of the $i_k(t)$, and so we can calculate ensemble averages such as $\langle i_k(t)\rangle$, $\langle i_k^2(t)\rangle$, $\langle i_k(t)i_l(t)\rangle$ or $\langle i_k(t)i_l(t+\tau)\rangle$ and use them to obtain, for example, $\langle I^2(t)\rangle$ or more complicated functions of $I(t)$. Two such quantities are of especial importance. The first is the correlation function, which is the ensemble average

$$\psi(\tau) = \left\langle \lim_{T\to\infty} \frac{1}{T}\int_0^T I(t)I(t+\tau)\,\mathrm{d}t \right\rangle, \tag{3.18}$$

and the second is MacDonald's function $\chi(\tau)$. This defined as follows.

We let

$$\overline{I^\tau} = \frac{1}{\tau}\int_t^{t+\tau} I(t)\,\mathrm{d}t$$

be the average of $I(t)$ over a finite period τ, $\bar{I}$ be the long-term average, which we also assume to be equal to the ensemble average, and $\overline{\Delta I^\tau} = \overline{I^\tau} - \bar{I}$. We then form the long-term average of $(\overline{\Delta I^\tau})^2$, which is $\overline{(\overline{\Delta I^\tau})^2}$, and finally $\chi(\tau)$ is defined in terms of the ensemble average of this expression, so that

$$\chi(\tau) = \tau^2\langle\overline{(\overline{\Delta I^\tau})^2}\rangle. \tag{3.19}$$

For example, for a simple shot-noise process this yields $\chi = eI_0\tau$. Both $\psi(\tau)$ and $\chi(\tau)$, in principle at least, are calculable from the physical structure of the system and the laws of statistical mechanics. A theorem due to Wiener (1930) and Khintchine (1934) then yields $w(f)$ in terms of $\psi(\tau)$ as

$$w(f) = 4\int_0^\infty \psi(\tau)\cos 2\pi f\tau\,\mathrm{d}\tau, \tag{3.20}$$

and a theorem due to MacDonald (1949) (see also Robinson (1958a)) yields it terms of $\chi(\tau)$ as

$$w(f) = 4\pi f\int_0^\infty \frac{\mathrm{d}\chi}{\mathrm{d}\tau}\sin 2\pi f\tau\,\mathrm{d}\tau. \tag{3.21}$$

To prove the Wiener–Khintchine theorem we express $I(t)$ during an indefinitely long interval T as

$$I(t) = \sum_n c_n \cos\left(\frac{2\pi nt}{T} + \phi_n\right), \tag{3.22a}$$

so that

$$I(t)I(t+\tau) = \sum_m \sum_n c_m c_n \cos\left(\frac{2\pi m t}{T} + \phi_m\right)\cos\left(\frac{2\pi n(t+\tau)}{T} + \phi_n\right). \quad (3.22b)$$

This gives

$$\lim_{T\to\infty} \frac{1}{T}\int_0^T I(t)I(t+\tau)\,\mathrm{d}t = \frac{1}{2}\sum_n c_n^2 \cos\frac{2\pi n\tau}{T},$$

and, since

$$w(f) = \lim_{T\to\infty} \frac{T}{2}\langle c_n^2\rangle,$$

we have

$$\psi(\tau) = \int_0^\infty w(f)\cos(2\pi f\tau)\,\mathrm{d}\tau. \quad (3.23)$$

Now $w(f)$ is undefined for $f < 0$, and so we can put $w(-f) = w(f)$, expressing (3.23) as

$$\psi(\tau) = \frac{1}{2}\int_{-\infty}^{\infty} w(f)\cos 2\pi f\tau\,\mathrm{d}f.$$

Since $\psi(\tau)$ is an even function of τ we can replace the lower limit in the Fourier inversion by zero and obtain the Wiener–Khintchine result (3.20).

To prove MacDonald's theorem we use the same expansion (3.22a) and this yields $\bar{I} = c_0$ and

$$\tau\,\overline{\Delta I^\tau} = 2\sum_{n=1}^{\infty} \frac{c_n T}{2\pi n}\cos\left(\frac{2\pi n(t+\tau/2)}{T} + \phi_n\right)\sin\frac{\pi n\tau}{T}.$$

The square of this leads to a double sum involving c_n and c_m, but a further average over T eliminates all terms except those with $n = m$, and so we obtain

$$\chi(\tau) = \tau^2\overline{\langle(\Delta I^\tau)^2\rangle} = 2\sum_{n=1}^{\infty}\left(\frac{T}{2\pi n}\right)^2\langle c_n^2\rangle\sin^2\frac{\pi n\tau}{T}.$$

This gives

$$\frac{\mathrm{d}\chi}{\mathrm{d}\tau} = \sum_{n=1}^{\infty}\frac{T}{2\pi n}\langle c_n^2\rangle\sin\frac{2\pi n\tau}{T},$$

and since $n/T = f$, we have, in terms of $w(f)$,

$$\frac{\mathrm{d}\chi}{\mathrm{d}\tau} = 2\int_0^\infty \frac{1}{2\pi f} w(f)\sin 2\pi f\tau\,\mathrm{d}f. \quad (3.24)$$

Again, if we set $w(-f) = w(f)$, this gives a Fourier integral

$$\frac{\mathrm{d}\chi}{\mathrm{d}\tau} = \int_{-\infty}^{\infty} \frac{1}{2\pi f} w(f)\sin 2\pi f\tau\,\mathrm{d}f,$$

and its inversion is

$$w(f) = 2\pi f \int_{-\infty}^{\infty} \frac{d\chi}{d\tau} \sin 2\pi f\tau \, d\tau = 4\pi f \int_{0}^{\infty} \frac{d\chi}{d\tau} \sin 2\pi f\tau \, d\tau,$$

since $d\chi/d\tau$ is clearly an odd function of τ. This is MacDonald's result.

Both these inversion formulae find applications in calculations of the noise power spectra due to identifiable physical processes in electronic components, and they can also be used to discuss the behaviour of signals and noise in complete electronic systems, especially when the signals are either present in digital form or occur in random sequences; from the point of view of this book, however, their chief significance is that they exist. They assure us that the power spectra that we use in our calculations can be related by a direct calculation to the physical processes occurring in particular devices. In fact, we shall not have to use them directly except in Chapter 5, where the Wiener–Khintchine theorem is used in the course of a discussion of the connection between noise and statistical mechanics. In the next section, however, we give an example of their use in connection with shot noise.

In the presence of coherent signals the power spectrum contains singularities, and then both the Wiener–Khintchine theorem and MacDonald's theorem require careful handling. Thus if $I(t) = A \cos 2\pi f_0 t$ we obtain

$$\psi(\tau) = \frac{A^2}{2} \cos 2\pi f_0 \tau,$$

and so

$$w(f) = 2A^2 \int_{0}^{\infty} \cos 2\pi f_0 \tau \cos 2\pi f\tau \, d\tau,$$

which, though clearly singular at $f = f_0$, is not otherwise well defined. The cumulative power spectrum is, however,

$$P(f) = \int_{0}^{f} w(f') \, df' = 2A^2 \int_{0}^{\infty} \int_{0}^{f} \cos 2\pi f_0 \tau \cos 2\pi f'\tau \, df' \, d\tau$$

and, integrated first over f, this gives

$$P(f) = \frac{A^2}{\pi} \int_{0}^{\infty} \frac{\cos 2\pi f_0 \tau \sin 2\pi f\tau}{\tau} \, d\tau$$

$$= \frac{A^2}{2\pi} \int_{0}^{\infty} \frac{(\sin 2\pi(f-f_0)\tau + \sin 2\pi(f+f_0)\tau) \, d\tau}{\tau} .$$

If $f > f_0$ both the terms in the integral are $\pi/2$, while if $f < f_0$ one term is $-\pi/2$ and the other $+\pi/2$ so that $P(f) = 0$ for $f < f_0$ and $P(f) = A^2/2$ for $f > f_0$, as we might expect.

MacDonald's function in this case is $\chi(\tau) = 2(A/2\pi f_0)^2 \sin^2 \pi f_0 \tau$ and so

$$w(f) = 4\pi f \int_0^\infty \frac{\mathrm{d}\chi}{\mathrm{d}\tau} \sin 2\pi f\tau \,\mathrm{d}\tau = 2\pi^2 \frac{f}{f_0} \int_0^\infty \sin 2\pi f_0\tau \sin 2\pi f\tau \,\mathrm{d}\tau.$$

Again, although $w(f)$ is not well defined by this expression it is easy to see that $P(f)$ increases discontinuously from zero below $f = f_0$ to $A^2/2$ above $f = f_0$.

3.3. The power spectrum of shot noise

We have already seen, in Chapter 2, that shot noise in a temperature-limited diode produces an effect on an amplifier of gain $G(f)$ as though it had a power spectrum $2eI_0$, where I_0 is the mean current. We now look at how this spectrum can be derived directly using first MacDonald's theorem and then the Wiener–Khintchine theorem.

For shot noise eqn (2.8) yields MacDonald's function $\chi(\tau)$ directly, as $\chi(\tau) = eI_0\tau$, and so

$$w(f) = 4\pi f \int_0^\infty \frac{\mathrm{d}\chi}{\mathrm{d}\tau} \sin 2\pi f\tau \,\mathrm{d}\tau = 4\pi f e I_0 \int_0^\infty \sin 2\pi f\tau \,\mathrm{d}\tau.$$

This is indeterminate but, if we replace the upper limit of the integral by T, we have

$$w(f) = 2eI_0(1 - \cos 2\pi fT),$$

and the cumulative power spectrum is

$$P(f) = 2eI_0 \int_0^f (1 - \cos 2\pi fT) \,\mathrm{d}f = 2eI_0\left(f - \frac{\sin 2\pi fT}{2\pi T}\right).$$

In the limit as $T \rightarrow \infty$ this gives $P(f) = 2eI_0 f$, and so the noise in a finite bandwidth $\mathrm{d}f$ is $2eI_0\,\mathrm{d}f$ and the power spectrum is effectively $w(f) = 2eI_0$.

From the Wiener–Khintchine theorem the correlation function for a simple shot-noise process in which transit time effects can be neglected is zero for $|\tau| > 0$ and singular at $\tau = 0$. To avoid discussing the nature of this singularity we shall consider a real physical situation, in which electrons crossing the diode take a finite time τ_0 to reach the anode. The current in the external circuit then consistes of pulses of current of duration τ_0 and total charge e. For simplicity we assume that the pulses are rectangular, so that the external current is a pulse of amplitude e/τ_0. This is not a good approximation in a thermionic diode but is a much better approximation for the important case of electrons traversing the depletion layer in a semiconductor diode.

If an electron emitted at t_k produces a current $eF(t-t_k)$ at time t then, in an interval T in which exactly m electrons are emitted, the anode current is

$$I_m(t) = e\sum_{k=1}^{m} F(t-t_k).$$

In this interval we have, averaging over the emission times t_k,

$$\overline{I_m(t)I_m(t+\tau)} = e^2\sum_{k=1}^{m}\sum_{l=1}^{m}\int_0^T \frac{\mathrm{d}t_1}{T}\cdots\int_0^T \frac{\mathrm{d}t_m}{T} F(t-t_k)F(t+\tau-t_l).$$

Just as we did in the case of Campbell's theorem, in § (2.3), we separate the terms with $k = l$ from those with $k \neq l$, replace the imits by $\pm\infty$, and obtain

$$\langle m\rangle - \langle m^2\rangle = \langle m\rangle^2 \quad \text{and} \quad e\langle m\rangle = I_0 T$$

The final ensemble average over the number of emitted electrons then yields, since $\langle m^2\rangle - \langle m\rangle = \langle m\rangle^2$ and $e\langle m\rangle = I_0 T$,

$$\psi(\tau) = \langle\overline{I_m(t)I_m(t+\tau)}\rangle = eI_0\int_{-\infty}^{\infty} F(t)F(t+\tau)\,\mathrm{d}t + I_0^2. \tag{3.25}$$

With $F(t) = 0$ for $t < 0$, $1/\tau$ for $0 < t < \tau_0$ and 0 for $t > \tau_0$ this gives

and

$$\left.\begin{aligned} \psi(\tau) &= eI_0\frac{\tau_0-\tau}{\tau_0^2} + I_0^2, & \tau < \tau_0 \\ \psi(\tau) &= I_0^2, & \tau > \tau_0 \end{aligned}\right\}. \tag{3.26}$$

The constant term I_0^2 gives a term

$$w_0 = 4I_0^2\int_0^{\infty} \cos 2\pi f_0\tau\,\mathrm{d}\tau,$$

which is indeterminate, but we have

$$P_0 = \int_0^f w_0\,\mathrm{d}f = \lim_{T\to\infty}\int_0^f\int_0^T 4I_0^2\cos 2\pi f\tau\,\mathrm{d}\tau\,\mathrm{d}f = \lim_{T\to\infty}\frac{2I_0^2}{\pi}\int_0^T \frac{\sin 2\pi f\tau\,\mathrm{d}\tau}{\tau}.$$

This changes discontinuously from $-I_0^2$ to $+I_0^2$ at $f = 0$ and is otherwise constant. It gives the d.c term $2I_0^2$ in the power spectrum. The remaining term is

$$w(f) = 4eI_0\int_0^{\tau_0} \frac{\tau_0-\tau}{\tau_0^2}\cos 2\pi f\tau\,\mathrm{d}\tau,$$

which gives

$$w(f) = 2eI_0\left\{\frac{\sin \pi f\tau_0}{\pi f\tau_0}\right\}^2.$$

At low frequencies, as the transit angle $2\pi f\tau_0$ tends to zero, this tends to $2eI_0$, but at high frequencies the noise power spectrum approaches zero. The effect of transit time is of some practical importance when temperature-limited thermionic diodes are used as standard noise sources.

3.4. Signal and noise in a continuous observation

In many cases, expecially when dealing with low-frequency signals such as those obtained from magnetic resonance spectrometers, the output of the system is recorded continuously, e.g. on a chart recorder, and is available for inspection at leisure. It is then of some interest to know the relation between the signal to noise ratio obtained by calculation from the extended record and the r.m.s. signal to r.m.s. noise ratio at the input to the recorder. Consider then a system whose output contains a number of coherent signal terms of sinusoidal form, and r.m.s. amplitudes S_n, and noise, of power spectrum $W(f)$, in a well-defined bandwidth Δf associated with the system. At the output of the system the instantaneous fluctuations have a mean square value $W(f)\Delta f$ (we assume for simplicity that the power spectrum is white so that W is independent of f at any rate within the bandwidth Δf), and so the signal to noise ratio for the nth component is $S_n/(W\,\Delta f)^{\frac{1}{2}}$. If the recorded output, including noise, is $V(t)$ it can be expressed in terms of a Fourier series $\sum_m \{a_m \cos(2\pi mt/T)+b_m \sin(2\pi mt/T)\}$, and the random part of the mth component has a mean square value $\langle a_m^2+b_m^2\rangle$, which is W/T since the frequency interval between components is $1/T$. The signal to noise ratio obtained from the record for the nth signal component is therefore $S_m/(W/T)^{\frac{1}{2}}$. Thus the record provides a signal to noise ratio improved in the ratio $(T\,\Delta f)^{\frac{1}{2}}$. Each component has the same uncertainty as we would have obtained by examining its instantaneous value using a system of bandwidth $1/T$ centred at the frequency of the signal component. Since an amplifier of bandwidth $1/T$ requires at least a time T to reach a steady state, the measurement, though apparently instantaneous, in practice takes as long to make. With the recorded output one observation period T can be used to examine many different signal components simultaneously present in the output of the system.

This example illustrates a rather general principle. If our access to a system extends back to a pair of terminals AB and across AB there exists a fluctuating voltage $V(t)$, which may or may not contain signals, then, if we record $V(t)$ over a period T and examine the result at leisure, using any *a priori* knowledge available to us about the nature of the signals, our final analysis will yield the optimum signal to noise ratio available in a time T. No process of filtering, correlating, or what have you, taking place in real time between the terminals AB and the final permanent record can produce a better signal to noise ratio than a leisured analysis of the complete record. It may be more convenient, but also it may eliminate signals potentially of interest.

3.5. Oscillographic presentation of a periodic signal

If the output of a system contains, in addition to noise, a periodic signal with components at f_0, $2f_0 \ldots$, etc., then the most natural presentation is on an oscilloscope with a repetitive time base at f_0. Presentations of this kind are used in radar, resonance spectrometers and many other applications. The resulting trace will be a stationary representation of a complete cycle of the signal together with a fluctuation due to noise. If we neglect the persistence of the screen phosphor successive traces will differ from each other, and if we fix our attention on one part of the trace, say the top of a pulse, the deflection at this point will fluctuate from trace to trace. If we denote this deflection by y and let the peak signal output be V, the noise power spectrum W, and the bandwidth Δf, then we shall obtain from a series of traces

$$\langle y \rangle = V, \tag{3.28a}$$

$$\langle \Delta y^2 \rangle = W\,\Delta f. \tag{3.28b}$$

Now suppose that the screen phospor has an exponential decay of luminosity with a time constant $1/\alpha$, so that I_0, the immediate light intensity, has decayed to $I_0 \exp(-n\alpha/f_0)$ after n cycles. If we suppose that in viewing the screen we tend to look at the centre of gravity of the light pattern, then after n cycles have passed this will be at a position

$$\bar{y} = \frac{\sum\limits_{k=0}^{n} y_k \exp(-k\alpha/f_0)}{\sum \exp(-k\alpha/f_0)},$$

where y_k was the actual deflection k periods before we made the observation.

The expectation value $\langle \bar{y} \rangle$ of the position of this centre of gravity is

$$\langle \bar{y} \rangle = \frac{\sum \langle y_k \rangle \exp(-k\alpha/f_0)}{\sum \exp(-k\alpha/f_0)} = \langle y \rangle, \tag{3.29}$$

since $\langle y_k \rangle$ does not depend on k. Thus the centre of gravity corresponds to the mean reading.

The fluctuations in the centre of gravity are

$$\langle \Delta \bar{y}^2 \rangle = \langle (\bar{y} - \langle y \rangle)^2 \rangle = \left\langle \left\{ \frac{\sum\limits_{0}^{n} \Delta y_k \exp(-k\alpha/f_0)}{\sum\limits_{0}^{n} \exp(-k\alpha/f_0)} \right\}^2 \right\rangle,$$

where Δy_k is the fluctuation in the kth trace. To simplify the arithmetic we assume that $n \rightarrow \infty$, i.e. $\exp(-n\alpha/f_0) \rightarrow 0$, and we then have

$$\langle \Delta \bar{y}^2 \rangle = \left\{1 - \exp(-\alpha/f_0)\right\}^2 \left\langle \sum_{k=0}^{\infty} \sum_{l=0}^{\infty} \Delta y_k \, \Delta y_l \exp\{-(k+l)\alpha/f_0\} \right\rangle.$$

Cross-terms with $k \neq l$ give zero and terms with $k = l$ give $\langle \Delta y_k^2 \rangle \exp(-2k\alpha/f_0)$. Since $\langle \Delta y_k^2 \rangle = \langle \Delta y^2 \rangle$, the sum is only over the exponential and we obtain

$$\langle \Delta \bar{y}^2 \rangle = \langle \Delta y^2 \rangle \frac{\cosh(\alpha/f_0) - 1}{\sinh(\alpha/f_0)}. \tag{3.30a}$$

If α/f_0 is large, i.e. the phosphor luminescence disappears between cycles, then the expression on the right is unity, but if α/f_0 is small, corresponding to a long decay time, we find

$$\langle \Delta \bar{y}^2 \rangle = \langle \Delta y^2 \rangle \frac{\alpha}{2f_0} = \frac{\alpha}{2f_0} W \, \Delta f. \tag{3.30b}$$

The effective noise bandwidth is reduced therefore from Δf to $\Delta f \, \alpha/2f_0$. If, for example, $\alpha = 10 \text{ s}^{-1}$ and $f_0 = 10^4$ Hz, the reduction is 2000-fold. This large increase in sensitivity will be achieved only if the signal does not change during a time $\sim 1/\alpha$. We should perhaps remark that even when the screen-phosphor decay is rapid, persistence of vision produces a similar effect.

The assumption, leading to eqn (3.30), that cross terms with $k \neq l$ give zero is valid in most practical cases. It is, however, worth noting that it is not valid if the input to the oscilloscope has been passed through a device such as a comb filter, which only transmits signal and noise components in a series of very narrow channels centred on the time-base repetition frequency and its harmonics.

Having shown that the description of noise by its power spectrum is equivalent to a statistical description we can now discuss the various sources of noise and their behaviour in electronic apparatus entirely in terms of power spectra.

3.6. Correlation power spectra

In later chapters of the book we shall show that noise in a number of important active devices, e.g. the transistor, can be represented by incorporating two noise generators, usually a voltage source and a current source in the equivalent circuit of the device. Often these two sources will be uncorrelated, but in some cases, where they derive entirely or in part from the same fundamental random process, they will be correlated. We need some way of representing the correlation between their Fourier components as a power spectrum. Perhaps it may be helpful to remark at this point that, in noise theory, the term power spectrum is used in the general sense of a spectral density associated with any second-order product of two fluctuating variables.

For example, it may correspond to the square of single variables (this has been our only use of the term so far), to the product of two different currents in different parts of a system, or to the product of a current and a voltage. Consider then two random processes leading to a noise current $I(t)$ and a noise voltage $V(t)$ and suppose that in a long interval T these have been expressed as Fourier series

$$I(t) = I_0 + \sum_{n=1}^{\infty}\left\{I_n \exp\left(\mathrm{j}\frac{2\pi nt}{T}\right) + I_n^* \exp\left(-\mathrm{j}\frac{2\pi nt}{T}\right)\right\}$$

and

$$V(t) = V_0 + \sum_{n=1}^{\infty}\left\{V_n \exp\left(\mathrm{j}\frac{2\pi nt}{T}\right) + V_n^* \exp\left(-\mathrm{j}\frac{2\pi nt}{T}\right)\right\}.$$

The product $I(t)V(t)$ will consist of a term I_0V_0, a sum of terms of the form $I_0V_n \exp(\mathrm{j}2\pi nt/T)$, etc. and a double sum of terms such as $I_nV_m \exp\{\mathrm{j} \times 2\pi(n+m)t/T\}$, $I_nV_m^* \exp \mathrm{j}\ 2\pi(n-m)t/T$, etc. When we form the ensemble average $\langle I(t)V(t)\rangle$. whatever the relation between I and V (i.e. even if $I(t) \equiv V(t)$) these terms will, because the phases are random, all average to zero, except for I_0V_0 and cross-terms such as $I_nV_n^*$ and $I_n^*V_n$. For $\langle I^2(t)\rangle$ we have

$$\langle I^2\rangle = I_0^2 + \sum_n \langle I_nI_n^* + I_n^*I_n\rangle,$$

and, as we showed in § 3.1, an appropriate definition of the power spectrum is

$$w_I(f) = \lim_{T\to\infty} 2T\langle I_n^*I_n\rangle, \tag{3.31}$$

with $f = n/T$. An appropriate definition for the correlation power spectrum related to the product $I(t)V(t)$ is therefore

$$w_{IV} = \lim_{T\to\infty} T\langle I_n^*V_n\rangle. \tag{3.32}$$

Clearly

$$w_{VI} = w_{IV}^* = \lim_{T\to\infty} T\langle V_n^*I_n\rangle, \tag{3.33}$$

and the contribution to $\langle I(t)V(t)\rangle$ from Fourier components in $\mathrm{d}f$ at f is

$$\mathrm{d}\langle I(t)V(t)\rangle = (w_{VI}+w_{IV})\,\mathrm{d}f \tag{3.34}$$

It is clear that, because of eqn (3.33), only one of the quantities w_{IV} or w_{VI} is required to define the correlation.

If $A(t)$, $B(t)$, and $C(t)$ are three random processes and $C(t) \equiv A(t)+B(t)$, the power spectra clearly satisfy the relation

$$w_C = w_{A+B} = w_A + w_B + w_{AB} + w_{BA}. \tag{3.35}$$

If, in addition, $B(t)$ is causally related to $A(t)$, so that a particular Fourier component of $B(t)$ at $f = \omega/2\pi$ can be expressed as $B(\omega) = \theta(\omega)A(\omega)$, where

$\theta(\omega)$ is a complex constant, we have:

$$w_{AB} = \theta(\omega)w_A \tag{3.36a}$$

$$w_B = \theta(\omega)\theta^*(\omega)w_A \tag{3.36b}$$

and

$$w_C = w_A(1+\theta+\theta^*+\theta\theta^*). \tag{3.37}$$

Finally we note that, if $\psi_{AB}(\tau)$ is defined as $\overline{A(t)B(t+\tau)}$ and $\chi_{AB}(\tau)$ as $\tau^2\langle\overline{A(t)B(t)^\tau}\rangle$, there are inversion relations analogous to the Wiener–Khintchine theorem and MacDonald's theorem which relate ψ and χ to w_{AB}.

4 Johnson Noise

EINSTEIN (1906) predicted that Brownian motion of the charge carriers would lead to a fluctuating e.m.f. across the ends of any resistance in thermal equilibrium. The effect was first observed by Johnson (1928), and its power spectrum was calculated by Nyquist (1928), whose treatment we now reproduce.

4.1. Nyquist's treatment

We consider two equal resistances R_1 and R_2, both at the same temperature T, connected in parallel. If the open-circuit fluctuating e.m.f. generated by R_1 has a mean square value $\langle V_1^2 \rangle$ then the power delivered by R_1 to R_2 is $\frac{1}{4}\langle V_1^2 \rangle / R$. If the resistances are to remain in thermal equilibrium this must equal the power delivered by R_2 to R_1. We therefore conclude that the fluctuating

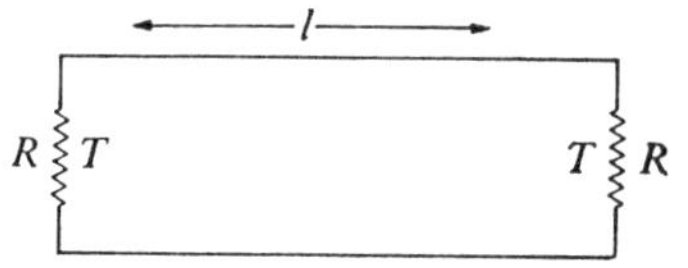

FIG. 4.1. Transmission line terminated at both ends by its characteristic impedance R.

e.m.f.s generated by two equal resistances at the same temperature are equal irrespective of the nature of the resistances. Furthermore, this equality must hold for each frequency component of the fluctuations; for, if in some range $\langle V_1^2 \rangle$ is greater than $\langle V_2^2 \rangle$, we could upset the over-all balance by removing that range by a filter containing only inductances and capacitances. Thus the power spectrum of the voltage fluctuations is a universal function of R, f, and T. To discover this function we consider the two resistances connected across either end of a long transmission line of characteristic impedance equal to R, length l, and wave velocity c (Fig. 4.1). If $w(f)$ is the power spectrum of the open-circuit voltage fluctuations, the mean power delivered by one resistance to the line in a frequency interval $\mathrm{d}f$ is

$$\mathrm{d}P = \frac{1}{4R} w(f)\,\mathrm{d}f. \tag{4.1}$$

After a time of the order of l/c a state of equilibrium is reached in which a mean power $\mathrm{d}P$ given by eqn (4.1) is flowing from left to right and an equal power from right to left. The energy in the line associated with $\mathrm{d}f$ is therefore

$$\mathrm{d}U = \frac{2l}{c}\,\mathrm{d}P = \frac{l}{2cR}w(f)\,\mathrm{d}f. \tag{4.2}$$

If the line is now shorted at both ends this energy is trapped as standing waves. The line is still in equilibrium at T since, if we connect it to R, no net power transfer will occur to or from the line. Therefore we can consider the line according to the usual methods of statistical thermodynamics and ascribe to each normal mode of the line of frequency f an energy

$$\epsilon = \frac{hf}{\exp\left(\dfrac{hf}{kT}\right)-1} \sim kT \quad \text{if} \quad hf \ll kT.$$

The normal modes of the shorted line are given by

$$f = nc/2l,$$

where n is an integer. Thus if l is large the number of modes in $\mathrm{d}f$ is

$$\mathrm{d}n = \frac{2l}{c}\,\mathrm{d}f,$$

and the mean thermal energy in $\mathrm{d}f$ is

$$\mathrm{d}U = \frac{2l}{c}kT\,\mathrm{d}f. \tag{4.3}$$

Equating our two expressions for $\mathrm{d}U$ we obtain

$$w(f) = 4RkT. \tag{4.4}$$

The spectrum of Johnson noise, like that of shot noise, is 'white', i.e. independent of f, and across any resistance R at T there exists a noise voltage of mean square value in a band Δf;

$$\langle V_n^2\rangle = 4kTR\,\Delta f. \tag{4.5}$$

If $R = 10^6\ \Omega$, $\Delta f = 10^6$ Hz, and $T = 293$ K (room temperature), then the r.m.s. noise voltage is 127 μV.

The maximum power available from the resistance is obtained when it is connected to an equal load R and is clearly

$$\langle P_n\rangle = \langle V_n^2\rangle/4R = kT\,\Delta f. \tag{4.6}$$

If Δf is 1 MHz and $T = 293$ K, this is 4×10^{-15} W.

4.2. Complex impedances

We now consider a complex impedance $Z = R(f)+\mathrm{j}X(f)$; let its voltage power spectrum be $w(f)$, and let it be connected to a resistance R at the same temperature. The average power transfer in df must be zero, and so

$$4kTR\,\mathrm{d}f.\frac{R(f)}{(R+R(f))^2+X^2(f)}=w(f)\,\mathrm{d}f.\frac{R}{(R+R(f))^2+X^2(f)},$$

thus

$$w(f)=4kTR(f). \tag{4.7}$$

The noise voltage across a complex impedance depends only on its real part. Pure reactances with no loss do not generate noise. We might have expected this result, for a lossless reactance has no means of coming into equilibrium with its environment.

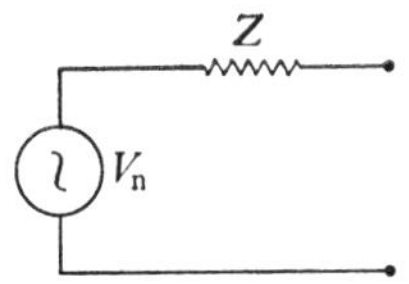

FIG. 4.2. Noise equivalent circuit of a passive impedance.

Fig. 4.2 therefore represents the equivalent circuit of an impedance $Z=R(f)+\mathrm{j}X(f)$ in equilibrium at T. The voltage generator has the power spectrum of eqn (4.7). It is sometimes more convenient to work with admittances and current generators. If $Y=G(f)+\mathrm{j}K(f)$ the equivalent circuit

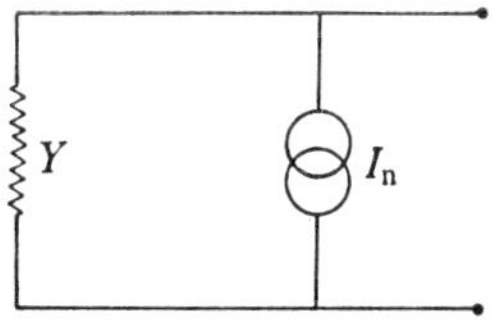

FIG. 4.3. Noise equivalent circuit of a passive admittance.

takes the form shown in Fig. 4.3, where the infinite impedance current generator has the power spectrum

$$w(f)=4kTG(f). \tag{4.8}$$

In either case the available noise power in df, i.e. the power delivered to a matched load $Z^*=R(f)-\mathrm{j}X(f)$ or $Y^*=G(f)-\mathrm{j}K(f)$, is just

$$\mathrm{d}P=kT\,\mathrm{d}f. \tag{4.9}$$

4.3. Generalizations

Because these results are based only on thermodynamic reasoning they apply to any system in thermal equilibrium whose electrical properties can be specified by a complex impedance or admittance. In particular, they apply to microphones, galvanometers, and aerials in constant-temperature enclosures. This last result has been used by Burgess (1941) to show that Nyquist's formulae lead to conclusions in accord with the laws of thermal radiation. It is interesting to note that Lorentz as long ago as 1906 used a very similar argument to derive the Rayleigh–Jeans law (Lorentz, 1952).

Johnson noise is of so fundamental an origin that it is frequently used as a yardstick with which to compare other sorts of noise. Thus, any noise source with a white voltage power spectrum w can be represented as Johnson noise in a resistance R_n at T with $4kTR_n = w$. If an actual physical resistance R is associated with the source we can take $R_n = R$ and adjust T accordingly. In other cases it is more convenient to take T as $T_0 = 293$ K (or room temperature) and to adjust R_n.

4.4. Noise figure

If we have a signal source of impedance R_s at T_s and available power P_s connected to a noiseless amplifier of bandwidth Δf, the output signal to noise ratio will be

$$\left(\frac{S}{N}\right)^2 = \frac{4P_s R_s}{4kT_s R_s \Delta f} = \frac{P_s}{kT_s \Delta f}, \tag{4.10}$$

which is independent of R and therefore of lossless transformations of the source impedance.

Any actual amplifier itself will introduce noise, and so the actual output signal to noise power ratio will be less than that given by eqn (4.10). Following Friis (1944) we may define a figure of merit for the amplifier, the noise figure F, by

$$F = \frac{(S/N)^2_{\text{input}}}{(S/N)^2_{\text{output}}}, \tag{4.11}$$

where $(S/N)^2_{\text{input}}$ is the ratio given by eqn (4.10) in terms of some nominal useful bandwidth of the system. The noise figure has a perfectly definite meaning only for a source of a definite temperature matched in a specified way to a linear amplifier of well-defined bandwidth. We shall consider the noise figure of linear amplifiers at greater length in Chapter 12.

4.5. Examples

We now illustrate the use of Nyquist's results by considering two simple examples.

In Fig. 4.4 the loss R in the tuned circuit is at a temperature T; what are the voltage fluctuations across the circuit? The resistance R is shunted by an impedance Z, where

$$1/Z = 2\pi j f C + 1/2\pi j f L.$$

The voltage power spectrum is therefore

$$w(f) = 4kTR\left|\frac{Z}{R+Z}\right|^2 = \frac{4kTR}{1+Q^2(f/f_0 - f_0/f)^2},$$

where $(2\pi f_0)^2LC = 1$ and $Q = 2\pi f_0 CR$. The voltage fluctuations are

$$\langle \Delta V^2 \rangle = \int_0^\infty w(f)\,\mathrm{d}f = 4kTR\frac{\pi}{2}\frac{f_0}{Q} = \frac{kT}{C}$$

and are independent of R and L. The mean energy stored in C is

$$\tfrac{1}{2}C\langle V^2 \rangle = \tfrac{1}{2}kT.$$

An equal energy is stored in L, and so we obtain the well-known result (from which we started) that the mean thermal energy of an oscillator is kT.

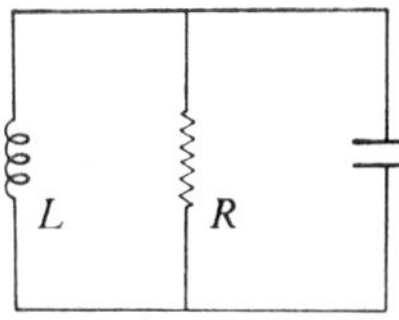

FIG. 4.4. Tuned circuit.

In Fig. 4.5 a transmission line of characteristic impedance R is correctly matched to a source of the same impedance. The source is at T_0 and the line, that attenuates a signal P to αP, is at T. What is the power spectrum of the voltage across the open end of the line? Suppose first that the line is matched

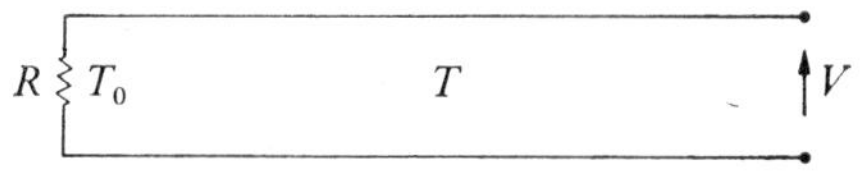

FIG. 4.5. Lossy line at T terminated by its characteristic impedance R at T_0.

at both ends to resistances R at the same temperature T. One resistance emits a power $kT\,\mathrm{d}f$ and receives a power $\alpha kT\,\mathrm{d}f$ from the other resistance. The balance $(1-\alpha)kT\,\mathrm{d}f$ must be contributed by the line. Thus in Fig. 4.5 the available power at the open end is $\alpha kT_0\,\mathrm{d}f+(1-\alpha)kT\,\mathrm{d}f$ and the voltage power spectrum is $4kR\{\alpha T_0+(1-\alpha)T\}$.

We conclude this chapter with a warning. Nyquist's results apply only to systems in internal thermal equilibrium. They do not apply directly to systems containing tubes, batteries, or temperature differences. Systems of that sort can only be treated, if at all, by separating them into their component parts, each of which is in thermal equilibrium, and then applying Nyquist's results to these separately.

5 Noise and Statistical Mechanics

5.1. Introduction

ANY macroscopic system, e.g. a resistor, a semiconductor diode, or a vacuum tube, contains an almost inconceivably large number of atomic or subatomic particles. A complete specification of the system in classical physics would involve a specification not only of the physical structure of the system but also of the coordinates and momenta of all these microscopic particles. In quantum mechanics the specification would have to be given in terms of an exact quantum state of the system as a whole and would involve the same number of parameters. In fact, for a macroscopic system we normally only have a macroscopic specification of its state in terms of a relatively small number of variables, and this specification is so coarse that it does no more than specify a large ensemble of possible microstates of the system. However, within this ensemble there may be sub-ensembles of microstates which, though consistent with the given macroscopic description, actually lead to different observable, or potentially observable, properties. Thus, if a small rod of carbon is described solely as a 1 MΩ resistor at a temperature of 300 K, there will be several sub-ensembles consistent with this description which differ from one another in the distribution of mobile charge in the rod. Some of these sub-ensembles correspond to states in which there is a macroscopically observable potential difference across the ends of the rod. If, for example, at some instant the potential difference is observed to be $+10\,\mu$V, then we know that, at that instant, the rod was in a state belonging to a particular sub-ensemble of states. However, because the rod, whose temperature is constant, must be in contact with other systems, or at least able to exchange energy between its various internal degrees of freedom, e.g. between charge-carriers and modes of lattice vibration, the rod will not persist in this particular sub-ensemble, and at a later instant is likely to be found exhibiting quite a different potential difference. On the other hand, if the two observations follow in close succession, the second potential difference is likely to be near the first in value. The correlation between two successive values will depend on both the interval between the measurements and the internal dynamics of the resistor. These determine how quickly one sub-ensemble situation disappears and a new situation arises. Obviously, in practice, the potential difference across the ends of the rod, being unpredictable, would be regarded as noise.

Thus we can regard noise and fluctuations as representing the random transitions of a system from one sub-ensemble to another, within an over-all ensemble of possible microstates limited only by the macroscopic description of the system. Provided that each sub-ensemble involved leads to observably different consequences, these fluctuations will be experimentally observable as noise.

In our example of a carbon rod or resistor, the over-all ensemble when the resistor is open circuited contains only microstates which lead to no net external macroscopic current. The fluctuations between sub-ensembles of this over-all ensemble can be observed, however, as fluctuations in charge distribution or potential difference. Conversely, if we consider a short-circuited rod, the over-all ensemble contains only microstates which lead to no potential difference, but now it also contains states corresponding to a net flow of current. In this case we shall observe fluctuations between sub-ensembles corresponding to different values of the current, and the observable noise will be current noise in the external lead.

The picture we have just developed is compatible with the description of macroscopic systems adopted in statistical mechanics. There, a macroscopic system is specified by giving its chemical composition and a limited number of macroscopic parameters, e.g. its temperature, its volume, and its state of strain. This determines an ensemble of possible microstates to which the system may belong. To each of these possible microstates there will correspond unique values for all possible macroscopic variables associated with the system, e.g. its total energy, its pressure, the voltage difference between two points in the system, and the current flowing in part of the system. Conversely, different values for each of these additional macroscopic variables define a group, or sub-ensemble, of microscopic states within the initial over-all statistical ensemble. The laws of statistical mechanics then allow us to calculate the probability of occurrence of each of these sub-ensembles and therefore of the particular values of the additional macroscopic variables. Furthermore, the internal structure of the system and the dynamical laws which govern its behaviour also allow us to calculate the probability that a system, known to be in one sub-ensemble at a time t_1, will have evolved to a second specified sub-ensemble at t_2. Thus if, for example, the two sub-ensembles are specified by the current I to which they correspond in an external lead, we can calculate the correlation function $\langle I(t_1)I(t_2)\rangle$. This, in conjunction with the Wiener–Khintchine theorem, is enough to allow us to obtain the noise power spectrum of the current I. We now use these concepts to discuss Johnson noise and shot noise and their connection with each other and the fundamental concepts of statistical mechanics. The reader unfamiliar with statistical mechanics may omit §§ (5.2) and (5.3) or alternatively refer to *Elementary statistical physics* by Kittel (1958), where the presentation of the basic laws is particularly relevant to this topic.

Statistical mechanics primarily furnishes a means of calculating the average properties of a macroscopic system, specified in enough macroscopic detail to determine a definite equilibrium state. It essentially achieves this by a calculation of the probability of occurrence of groups of microscopic states. If, for example, in a resistor of specified construction, we define a group of states for the conduction electrons as states in which the x component of the electron velocity is between v_x and v'_x, then statistical mechanics will furnish the probability that n electrons will be found in this group of states. From this probability we can calculate the mean value of the number of electrons in these states and also the fluctuations about this mean value. In other words, if in a group of states the mean value is N_i and in a succession of instantaneous observations we find $N_i+\nu_{i1}$, $N_i+\nu_{i2}$, $N_i+\nu_{i3}$, etc., then we can predict the mean square deviation

$$\langle \nu_i^2 \rangle = \lim_{M\to\infty} \frac{1}{M} \sum_{k=1}^{M} \nu_{ik}^2.$$

Statistical mechanics therefore makes direct statements about instantaneous fluctuations of occupation numbers from their mean values. Our task is to relate these statements to results for the temporal development of fluctuations in the macroscopic observables determined by the occupation numbers.

5.2. Fundamental formulae

We assume, for the sake of clarity and brevity, that we are dealing with a system such as a resistor, a semiconductor diode, or a vacuum tube equipped with two terminals joined by an external lead of negligible resistance and that the current I in this lead is the macroscopic variable of interest. The mobile charge carriers (which we shall refer to as electrons, though they may be ions or holes) in the system have a charge e, and the microscopic states of the system are classified in such a way that, if at time t there are $n_i(t)$ electrons in the group of states labelled by i, then the resulting contributions to the current I at t is $\zeta_i n_i(t)$, where ζ_i is a definite function calculable from the structure of the system. Thus the total current is the sum over groups of states

$$I(t) = \sum_i \zeta_i n_i(t). \tag{5.1}$$

Next we divide each n_i into two parts, a mean part N_i and a fluctuation $\nu_i(t)$. The mean current is therefore

$$\langle I \rangle = \sum_i \zeta_i N_i \tag{5.2}$$

and the fluctuation is $\sum_i \zeta_i \nu_i(t)$. We shall assume henceforth either that $\langle I \rangle$ is zero or that we are dealing only with the fluctuating component, i.e. that I

represents $I-\langle I\rangle$. The correlation function for I is therefore

$$\psi(t) = \langle I(t)I(0)\rangle = \sum_i \sum_j \zeta_i\zeta_j\langle \nu_i(t)\nu_j(0)\rangle, \tag{5.3}$$

where we have assumed that we are dealing with a stationary process so that the origin of time is irrelevant. The power spectrum of I is

$$w(f) = 4\int_0^\infty \psi(t)\cos 2\pi ft\, \mathrm{d}t = 4\sum_i\sum_j \zeta_i\zeta_j\mu_{ij}, \tag{5.4}$$

where

$$\mu_{ij} = \int_0^\infty \langle \nu_i(t)\nu_j(0)\rangle \cos 2\pi ft\, \mathrm{d}t. \tag{5.5}$$

The evaluation of μ_{ij} requires first a knowledge of the way fluctuations ν_i arise, and secondly a knowledge of the way they decay. The second requirement involves the detailed dynamics of the system and will be considered later, but the first can be related to the following fundamental result of statistical mechanics. If the mean occupation number of a group of states is N_i in a system at a temperature T, then the mean square value of the fluctuations from N_i at individual instants is given by

$$\langle \nu_i^2\rangle = kT\frac{\partial N_i}{\partial \mu}, \tag{5.6}$$

where k is Boltzmann's constant and μ is the chemical potential. For a system at constant temperature and pressure throughout, this is equivalent to

$$\langle \nu_i^2\rangle = -kT\frac{\partial N_i}{\partial E_i}, \tag{5.7}$$

where E_i is the energy per particle of the states in question. Furthermore, these results can be simplified further once we know whether the particles obey Fermi–Dirac, classical, or Bose–Einstein statistics. The results are

$$\langle \nu_i^2\rangle = n_i(1-n_i/g_i), \tag{5.8FD}$$

$$\langle \nu_i^2\rangle = n_i, \tag{5.8C}$$

$$\langle \nu_i^2\rangle = n_i(1+n_i/g_i), \tag{5.8BE}$$

where g_i is the total number of distinguishable microscopic states in the group labelled by i. In a non-degenerate situation, where $n_i \ll g_i$, the classical result is universally valid. We shall see the importance of this later, in connection with shot noise.

We now have all the basic results required, and in the next section we apply them to Johnson noise in a conductor. The treatment is quite general and applies whether or not the carriers are electrons, and whether or not the conductor is degenerate.

5.3. Johnson noise

In Fig. 5.1 we show a conductor of arbitrary shape with two terminals, one earthed and the other at a potential ϕ. This potential ϕ produces a field **F** at the point **r** and as a result a current of density **J**, not necessarily parallel

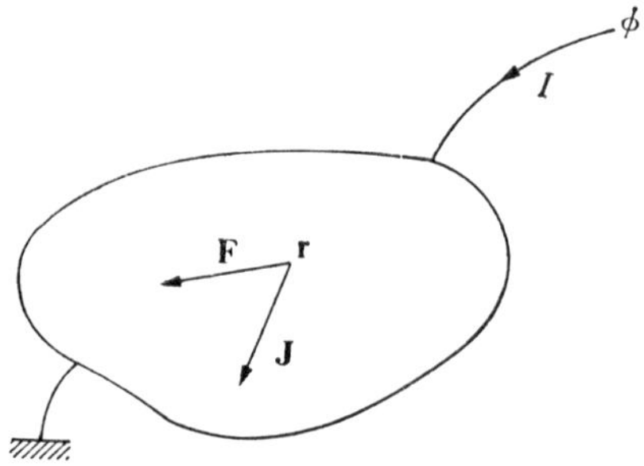

FIG. 5.1. A conductor with two terminals.

to **F**, flows in the conductor at **r**. However, we do assume that the system is linear, so that both **F** and **J** are proportional to ϕ. We write

$$\mathbf{F}(\mathbf{r}) = \phi\mathbf{S}(\mathbf{r}), \tag{5.9}$$

where **S** is a real vector function of position.

If E is the energy of the system, the rate P_i at which the field **F** does work on the system, as a result of the motion of the ith particle of charge e and momentum $\mathbf{p}_i$ at **r**, is

$$P_i = e\mathbf{F}(\mathbf{r})\cdot\nabla_i E = e\phi\mathbf{S}(\mathbf{r})\cdot\nabla_i E, \tag{5.10}$$

where $\nabla_i E$ is an abbreviated notation for $(\nabla_\mathbf{p} E)_{\mathbf{p}=\mathbf{p}_i}$ i.e. its x component is $(\partial E/\partial p_x)_{\mathbf{p}_i}$, and $\nabla_i E$ is the velocity of the ith particle.

In a conductor there is no change in the field energy of the system (see Appendix A for a fuller discussion) as a result of the motion of individual charge carriers, and therefore we can equate P_i to the power delivered by the source of the potential difference ϕ. Thus, as a result of the motion of this particular carrier, a current

$$I_i = e\mathbf{S}(\mathbf{r}).\nabla_i E \tag{5.11}$$

flows in the external lead. If we now classify the electrons, or carriers, by their positions **r** and momenta $\mathbf{p}_i$, the function ζ in eqn (5.1) which gives I in terms of the occupation numbers becomes a function of two parameters **r** and $\mathbf{p}_i$, as do the occupation numbers. We express this as

$$\zeta_{ir} = e\mathbf{S}(\mathbf{r}).\nabla_i E, \tag{5.12}$$

so that

$$I(t) = e\sum_i\sum_r \mathbf{S}(\mathbf{r}).\nabla_i E\,.\,n_{ir}(t). \tag{5.13}$$

The quantity μ_{ij} which appears in eqns (5.4) and (5.5) now becomes a function of four parameters: two momenta $\mathbf{p}_i$ and $\mathbf{p}_j$ and two coordinates $\mathbf{r}$ and $\mathbf{s}$. We shall not indicate this explicitly, but to indicate that in an expression such as, for example,

$$\psi(t) = e^2 \sum_{ir} \sum_{js} (\mathbf{S}.\nabla_i E)(\mathbf{S}.\nabla_j E)\langle v_i(t)v_j(0)\rangle,$$

one factor $\mathbf{S}$ depends on $\mathbf{r}$ paired with $\mathbf{p}_i$, and the other on $\mathbf{s}$ paired with $\mathbf{p}_j$, while $v_i(t)$ depends on $\mathbf{p}_i$ and $\mathbf{r}$, and $v_j(0)$ on $\mathbf{p}_j$ and $\mathbf{s}$, we shall place all the parameter labels on the summation signs.

We now have to evaluate

$$\mu_{ij} = \int_0^\infty \langle v_{ir}(t)v_{js}(0)\rangle \cos 2\pi ft \, dt, \tag{5.14}$$

but, before we attempt this, we shall look briefly at the Boltzmann transport equation, which determines the dynamical behaviour of the carriers. This equation is

$$\frac{dn_i}{dt} = \left(\frac{\partial n_i}{\partial t}\right)_{\text{collisions}}, \tag{5.15}$$

i.e. the total influx of particles into a cell or group of states during the motion is equal to the influx due to collisions. In the absence of an applied field (i.e. with the leads in Fig. 5.1 shorted) the mean occupation numbers do not change and so eqn (5.15) reduces to an equation of motion for the fluctuations which is

$$\frac{\partial v_i}{\partial t} = \left(\frac{\partial v_i}{\partial t}\right)_{\text{collisions}}. \tag{5.16}$$

The time derivative on the left is a partial derivative, since we are considering a group of states of fixed momentum and position. Because we are considering only small fluctuations we can assume that the collision term is linear in the fluctuations but, since fluctuations in one group influence those in a similar nearby group, we shall have a linear operator relation of the form

$$\frac{\partial v_i}{\partial t} = \sum_{js} C_{ij} v_j, \tag{5.17}$$

where the rows of the matrix C_{ij} are labelled by i and r and the columns by j and s.

We now return to eqn (5.14), which we integrate twice by parts. Since $\langle v_{ir}(\infty)v_{js}(0)\rangle = 0$ this yields

$$(2\pi f)^2 \mu_{ij} = -\langle \dot{v}_i(0)v_j(0)\rangle - \int_0^\infty \langle \ddot{v}_i(t)v_j(0)\rangle \cos 2\pi ft \, dt. \tag{5.18}$$

We now let $\omega = 2\pi f$ and use (5.17) to eliminate the time derivatives, and then, setting

$$u_{ij} = \langle v_i(0)v_j(0)\rangle, \tag{5.19}$$

we obtain

$$\omega^2\mu_{ij} = -C_{ik}u_{kj} - C_{ik}C_{kl}\mu_{lj}, \tag{5.20}$$

in which summation over repeated subscripts is implied. In matrix notation this equation is

$$(\omega^2+C^2)\mu = -Cu. \tag{5.21}$$

Now, although the matrix C has at least five zero eigenvalues, corresponding to the constants of the motion (number of particles, total energy and three components of the total momentum), the matrix ω^2+C^2 is positive definite and so has an inverse. If we let

$$D = (\omega^2+C^2)^{-1}C, \tag{5.22}$$

then the solution of (5.21) is

$$\mu = -Du. \tag{5.23}$$

In component notation, with the full array of state-labels as subscripts, this is

$$\mu_{ir,js} = -D_{ir,kt}u_{kt,js}. \tag{5.24}$$

This equation expresses the components μ_{ij} (which yield the power spectrum $w(f)$ according to eqn (5.4), with ζ_i replaced by (5.12)) in terms of the fluctuations u at a single instant and a matrix D which contains the dynamics of the system. This matrix describes the coupling between the charge carriers and the heat-bath provided by the lattice. Clearly it is connected with the resistance of the system, but first we consider a special case which clarifies its meaning. If all the fluctuations in every class have the same relaxation time τ, then $C = -1/\tau$ and so

$$D = -\frac{\tau}{1+\omega^2\tau^2}.$$

The quantity D describes how long a particular Fourier component of a fluctuation, at a frequency $\omega/2\pi$, persists after its first appearance.

The single-instant fluctuations u can be obtained from (5.7) and, since fluctuations (at an instant in time) of the occupation numbers of mutually exclusive groups of states are uncorrelated, we have

$$\mu_{ij} = -\delta_{ij}\delta_{rs}kT\frac{\partial N_{js}}{\partial E}. \tag{5.25}$$

Thus, finally, putting all these results together we obtain the power spectrum as

$$w = 4e^2kT\sum_i\sum_r\sum_j\sum_s(\mathbf{S}(\mathbf{r})\cdot\nabla_i E)(\mathbf{S}(\mathbf{s})\cdot\nabla_j E)D_{ir,js}\frac{\partial N_{js}}{\partial E}. \tag{5.26}$$

We now reduce this expression to a much simpler form.

In the presence of the field $\mathbf{F}$, due to the applied potential ϕ, the Boltzmann transport equation is

$$\frac{\partial \nu_{ir}}{\partial t}+e\mathbf{F}\cdot\nabla_{\mathbf{p}}(N_{ir}+\nu_{ir}) = \sum_{j}\sum_{s} C_{ir,js}\nu_{js}, \tag{5.27}$$

where now ν_{ir} is the small disturbance from the mean distribution N_{ij}, caused by the field $\mathbf{F}$. This may be solved to first order in $\mathbf{F}$, and if we assume that both $\mathbf{F}$ and ν vary as exp (jωt), we have

$$(C-\mathrm{j}\omega)\nu = e\mathbf{F}.\nabla_{\mathbf{p}}N,$$

so that

$$\nu = (C^2+\omega^2)^{-1}(C+\mathrm{j}\omega)e\mathbf{F}.\nabla_{\mathbf{p}}N. \tag{5.28}$$

The power dissipated near $\mathbf{r}$ due to carriers of momentum $\mathbf{p}_i$ is given by the real part of $\frac{1}{2}\nu_{ir}e\mathbf{F}^*.\nabla_i E$, which does not involve the imaginary term in (5.28). The remaining real part is the matrix D defined in (5.22), and so the total power dissipated is

$$P = \tfrac{1}{2}e^2\sum_i\sum_r\sum_j\sum_s (\mathbf{F}^*(\mathbf{r})\cdot\nabla_i E)\,D_{ir,js}(\mathbf{F}(\mathbf{s})\cdot\nabla_j N_{js}).$$

Now $\mathbf{F} = \phi\mathbf{S}$ and if $Y = \mathrm{j}X+G$ is the admittance measured at the terminals in Fig. (5.1) the power dissipated is $\frac{1}{2}\phi\phi^*G$, thus we have

$$G = e^2\sum_i\sum_r\sum_j\sum_s (\mathbf{S}(\mathbf{r})\cdot\nabla_i E)\,D_{ir,js}(\mathbf{S}(\mathbf{s})\cdot\nabla_j N_{js}). \tag{5.29}$$

But we also have

$$\nabla_j N_{js} = \frac{\partial N_{js}}{\partial E}\nabla_j E, \tag{5.30}$$

and so the sums in eqns (5.29) and (5.26) are identical except for a factor $4kT$. The power spectrum of the current in the external lead is therefore

$$w = 4kTG. \tag{5.31}$$

This (see eqn (4.8)) is Nyquist's result. The treatment in this section is certainly no more rigorous, and also certainly much more tedious, than Nyquist's; however, it does bring out how the fluctuations are related to the type of calculation which occurs in statistical mechanics. This is important, because there is a widespread belief that fluctuations as described in statistical mechanics are directly related to fluctuations observed as noise. The complexity of the relation should be obvious, at least, at this point.

We see then that the current fluctuation is a weighted sum of the cross-power spectra μ_{ij} of the population fluctuations. These cross-power spectra in turn depend (1) on the magnitude u_{ij} of the fluctuations which arise at any one instant and (2) on the dynamical process, described by the collision term in the Boltzmann equation, which leads to their decay. The first part (1) depends only on the equilibrium properties of the carrier distribution, while (2) depends on the coupling between the carriers and the lattice. In the final result (1) is summarized by the term kT and (2) by the conductance term G. The discussion depends critically on the assumption that the occupation number fluctuations which lead to the term u_{ij} are those characteristic of thermal equilibrium. As long as this is satisfied the power spectrum is bound to be given by (5.31). This formula therefore applies to any system in which the charge carriers are in equilibrium with a lattice, and can be assigned a temperature T equal to that of the lattice. We shall see the importance of this result when we come, in Chapter 11, to deal with noise in field-effect transistors. In the next section we apply similar arguments to shot noise, where we are dealing with carriers not in thermal equilibrium.

5.4. Shot noise

In vacuum tubes and semiconductor devices incorporating a p–n junction, the charge carriers interact with fields due to applied external voltages only in specific regions of the device: in one case in the vacuum space between the cathode and other electrodes, in the other case in the depletion layer at the junction. The reason for this is simply that the fields within the devices, due to applied voltages, are zero except in these regions. In a vacuum tube it is obvious that in the interaction region the electrons are no longer exchanging energy with a heat-reservoir or lattice; in a p–n junction this is less obvious and depends on the fact that in a depletion layer the motion of the electrons is dominated by the effect of the macroscopic field in the layer, and little influenced by the very small number of collisions (if any) that occur in the layer. Therefore, in both cases fluctuations in the behaviour of electrons in the interaction region are due to fluctuations in the emission of electrons into the region, i.e. to shot noise. Shot noise is simpler than thermal noise because electrons, once emitted, are effectively free from further fluctuating influences.

We divide electrons leaving the emitter region, or cathode, into mutually exclusive classes $(i, \mathbf{r})$, specified by their momenta $\mathbf{p}_i$ and position $\mathbf{r}$ as they cross the emitter boundary or cathode surface, and we consider the number $n_{\mathbf{r}iT}$ of electrons emitted in each class in a time T. First we divide T into K intervals, labelled by k, of length t, short compared with any relaxation or collision time in the emitter. Because no collisions occur in one of the intervals, all electrons emitted in a given class $(i, \mathbf{r})$ during the interval k came from a definite group of states at the beginning of this interval, and all electrons in this group were emitted during the interval. Electrons emitted in other classes during this same interval and also electrons emitted either in this class or in other classes during other intervals were not in this group of states at the beginning of the interval t. Since fluctuations in the occupation numbers of different groups of states at the same time are uncorrelated and, since electrons in the class $(i, \mathbf{r})$ leave the emitter during the interval k without collisions, there are no correlations either between fluctuations in different emitted classes or between fluctuations in the same class in different intervals. If, therefore, ν_{irt} is the fluctuation in the number of electrons in the class (i, r) emitted in one interval, the fluctuations in a finite interval T containing K elementary intervals satisfy both

$$\langle \nu_{irT}^2 \rangle = \sum_{k=1}^{K} \langle \nu_{irt}^2 \rangle \tag{5.32}$$

and

$$\langle \nu_{irT} \nu_{jsT} \rangle = 0 \quad \text{if } r \neq s \quad \text{or} \quad i \neq j. \tag{5.33}$$

The emitted electrons enter a region in which the statistics are non-degenerate or classical. It follows that either the distribution in the emitter was non-degenerate or that it was degenerate but the electrons had to surmount a potential barrier at the emitter surface in order to escape. The first case corresponds roughly to a p–n junction, the second to a thermionic cathode. In either case the fluctuations in the emitted classes of electrons are classical and so

$$\langle \nu_{irt}^2 \rangle = N_{irt}, \tag{5.34}$$

where N_{irt} is the mean number emitted in t. We now have

$$\langle \nu_{irT}^2 \rangle = N_{irT}. \tag{5.35}$$

and the current fluctuations are

$$\langle (\overline{\Delta I^r})^2 \rangle = \frac{e^2}{T^2} \langle \nu_{irT}^2 \rangle = \frac{e}{T} \frac{eN_{irT}}{T} = \frac{e}{T} I_{0ir}, \tag{5.36}$$

where I_{0ir} is the mean emission current in class (i, r). MacDonald's function $\chi_{ir}(T)$ for each class is therefore eTI_{0ir} and (see Chapter 3) each class contributes

$$w_{ir} = 2eI_{0ir} \tag{5.37}$$

to the noise power spectrum of the current. The total power spectrum for the whole current is

$$w = \sum_{ir} w_{ir} = 2e \sum_{ir} I_{0ir} = 2eI_0. \tag{5.38}$$

This coincides with our earlier formula, but we have some additional information contained in (5.37). If the emitted current is divided arbitrarily into exclusive classes, e.g. by its position on the emitter surface $\mathbf{r}$ or the electron momentum $\mathbf{p}_i$ on emission, then each class individually displays the typical shot-noise power spectrum, and fluctuations in different classes are uncorrelated. We shall explore some of the consequences of this in the next section, and also use the result later in the book.

We are now able to state the distinction between a shot-noise process and a thermal-noise process. In a shot-noise process the region in which the charge carriers interact with external fields is physically distinct from the region in which their statistical properties are established. Thus, as they interact with external fields, they neither influence nor are influenced by random processes occurring in the emitter region. In a thermal-noise process, on the other hand, the interaction region coincides with the region where the carrier fluctuations are generated and, during their interaction, the carriers remain in approximate thermal equilibrium with a heat-reservoir or lattice. In this interaction region, fluctuations in occupation numbers arise and decay continuously as the carriers interact via collisions with the lattice. The amplitude of the fluctuations depends on kT and the dynamics of their decay on the nature of the collision processes, which also determines the conductance G.

Thus, in a bipolar transistor, the fluctuations in the active carrier flow are established within the emitter and, to a lesser extent, within the base where collisions occur but there is no field due to externally applied voltages. Fields and interactions occur only in the emitter–base and base–collector depletion layers. Bipolar transistors are typical shot-noise limited devices. By contrast, in a field-effect transistor the fluctuations are established in the channel, and this is also where the carriers interact with the applied voltages; a field-effect transistor is therefore a thermal-noise limited device.

5.5. Shot noise, velocity noise, and emitter temperature

The over-all power spectrum (5.38) for the total emitted current makes no explicit reference to the temperature of the emitter but, if we consider the individual components in (5.37), the temperature of the emitter influences the partition of the total emitted current into classes, associated with momenta $\mathbf{p}_i$. The higher the value of the emitter temperature the higher will be the values of p_i which make a substantial contribution to the total current. This has important consequences both in vacuum tubes and in bipolar semiconductor devices, which we now discuss.

In vacuum tubes, the gain mechanism is substantially independent of the thermal spread in the velocities of emission of electrons from the cathode. On the other hand, the electrons in the active region interact through long-range Coulomb forces, and this leads to correlations between fluctuations associated with classes of electrons which had different thermal velocities on emission. The detailed mechanism is exceedingly complicated and rather different, depending on whether we are dealing with a space-charge limited device such as a triode at low frequencies, where transit time can be neglected, or an electron-beam tube such as a travelling-wave tube, where transit time is all important. In the case of the triode the correlations tend to reduce the over-all power spectrum of the current from $2eI_0$ to $\Gamma^2 2eI_0$, where the smoothing factor Γ^2 depends on the ratio of the mean thermal energy of the electrons in the grid-to-cathode region to their coherent or d.c. energy. The larger the range of thermal emission velocities which contribute to I_0 the less effective is this smoothing process and, in general, Γ^2 is of the order of kT/eV_e, where T is the cathode (emitter) temperature and eV_e is the d.c.

energy of electrons near the grid. In a triode, therefore, the signal properties are independent of T but the mean square output noise increases with T.

In a microwave-beam tube the electrons interact with external fields in a drift region well away from the cathode, and the noise which appears in the external circuit arises partly from current fluctuations in the beam at the input to the drift region and partly from fluctuations in the velocity of electrons entering the drift region. These fluctuations, as we shall see, depend on T. Thus again the noise to signal ratio in the output of the device increases with T.

In a bipolar transistor the noise in the output (collector current) has a power spectrum given by (5.38) and is independent of T. The signal amplifying process depends, however, on the mutual conductance or rate of variation of collector and emitter current with the applied base–emitter voltage. This process operates by altering the range of electrons which can be emitted, thus, as this range increases with increasing T (emitter temperature), the fraction of the total emitter current that changes in response to a given change in base–emitter voltage decreases. The signal output, for a given input, decreases with increasing T. Thus, once again, the output noise to signal ratio increases with T. In every case the output noise to signal ratio increases with temperature of the region in which the statistical properties of the active charge carriers are determined, and this appears to be quite general. We shall see several examples of this rule in later chapters. It applies to all types of amplifying tube and transistor and also, in a slightly different sense, to parametric amplifiers and masers. The one exception is the photodetector, where the active charge carriers (photo-electrons) have statistical properties determined not by a thermal equilibrium process but by an intrinsically non-equilibrium photo-electric interaction.

We now look more quantitatively at the concept of velocity fluctuations in a stream of emitted electrons, a notion originally due to Rack (1938).

If, in any period T, n_i electrons of velocity v_i cross a fixed plane in space, the average velocity of these electrons is

$$\overline{v^T} = \frac{\sum_i n_i v_i}{\sum_i n_i}.$$

Now suppose that the average number of electrons crossing in a period T with velocity v_i is N_i, so that the long-term average velocity is

$$\bar{v} = \frac{\sum_i N_i v_i}{\sum_i N_i},$$

then, in one period T, the fluctuation in the average velocity is

$$\overline{\Delta v^T} = \overline{v^T} - \bar{v} = \frac{\sum n_i v_i}{\sum n_i} - \frac{\sum N_i v_i}{\sum N_i}.$$

If we let $\nu_i = n_i - N_i$ be the fluctuations in the numbers n_i and assume that $|\sum \nu_i| \ll \sum N_i$, we have

$$\overline{\Delta v^T} \approx \frac{1}{\sum N_i}\left\{\sum \nu_i v_i - \sum N_i v_i\left(\frac{\sum \nu_i}{\sum N_i}\right)\right\},$$

which we can also express as

$$\overline{\Delta v^T} = \frac{1}{\sum N_i}\left\{\sum \nu_i v_i - \bar{v}\sum \nu_i\right\}. \tag{5.39}$$

Because fluctuations in different velocity classes are uncorrelated the ensemble average of $(\overline{\Delta v^T})^2$ is

$$\langle(\overline{\Delta v^T})^2\rangle = \left(\frac{1}{\sum N_i}\right)^2\left\{\sum \langle \nu_i^2\rangle v_i^2 - \bar{v}^2 \sum \langle \nu_i^2\rangle\right\},$$

and further we have $\langle \nu_i^2\rangle = N_i$ so that

$$\langle(\overline{\Delta v^T})^2\rangle = \frac{1}{\sum N_i}\{\overline{v^2} - (\bar{v})^2\}. \tag{5.40}$$

If I_0 is the total current we have $e\sum N_i = I_0 T$, and so

$$\langle(\overline{\Delta v^T})^2\rangle = \frac{e}{I_0 T}\{\overline{v^2} - (\bar{v})^2\}. \tag{5.41}$$

MacDonald's function $\chi(T)$ is therefore

$$\chi(T) = T^2\langle(\overline{\Delta v^T})^2\rangle = \frac{eT}{I_0}\{\overline{v^2} - (\bar{v})^2\}.$$

which gives a power spectrum for the velocity fluctuations at the plane in question equal to

$$w_v = \frac{2e}{I_0}\{\overline{v^2} - (\bar{v})^2\}. \tag{5.42}$$

This is Rack's result.

The current fluctuation is obviously

$$\overline{\Delta I^T} = \frac{e}{T}\sum \nu_i$$

and the power spectrum of I is clearly

$$w_I(f) = 2eI_0,$$

but we can also calculate the cross-correlation

$$\langle \overline{\Delta I^T}\,\overline{\Delta v^T}\rangle = \left\langle \frac{e}{T\sum N_i}\left(\sum \nu_i v_i - \bar{v}\sum \nu_i\right)\sum \nu_j\right\rangle$$

and, again because fluctuations in different classes are uncorrelated, this reduces to

$$\langle \overline{\Delta I^T}\,\overline{\Delta v^T}\rangle = \frac{e}{T\sum N_i}\Big(\sum v_i v_i^2 - \bar{v}\sum v_i^2\Big) = \frac{e}{T\sum N_i}\Big(\sum v_i N_i - \bar{v}\sum N_i\Big) = 0.$$

Thus the velocity and current fluctuations are uncorrelated, provided of course that no coherent interaction has taken place between the surface of the emitter (cathode) and the plane at which we evaluate the fluctuations.

At the surface of the cathode, where the electrons have a Boltzmann distribution of velocities normal to the surface, we have, in terms of the cathode temperature T,

$$\bar{v} = \int_0^\infty v\exp\Big(-\frac{mv^2}{2kT}\Big)\,\mathrm{d}\Big(\frac{mv^2}{2kT}\Big) = \Big(\frac{\pi kT}{2m}\Big)^{\frac{1}{2}}$$

$$\overline{v^2} = \frac{2kT}{m},$$

and so

$$w_v(f) = \frac{e}{m}\frac{4kT}{I_0}\Big(1-\frac{\pi}{4}\Big). \tag{5.43}$$

We can also evaluate $\bar{v}$, $\overline{v^2}$, and w_v at a plane at a positive potential ϕ relative to the cathode, thus, for example,

$$\bar{v} = \int_0^\infty \Big(v_1^2+\frac{2e\phi}{m}\Big)^{\frac{1}{2}}\exp\Big(-\frac{mv_1^2}{2kT}\Big)\,\mathrm{d}\Big(\frac{mv_1^2}{2kT}\Big)$$

and, setting $\alpha = e\phi/kT$, we obtain

$$\overline{v^2}-(\bar{v})^2 = \frac{2kT}{m}f(\alpha) \tag{5.44}$$

and

$$w_v = \frac{e}{m}\frac{4kT}{I_0}f(\alpha), \tag{5.45}$$

where

$$f(\alpha) = 1+\alpha-\{\alpha^{\frac{1}{2}}+\tfrac{1}{2}\pi^{\frac{1}{2}}\mathrm{e}^{\alpha}(1-\operatorname{erf}\alpha^{\frac{1}{2}})\}^2 \tag{5.46}$$

with

$$\operatorname{erf}\alpha = 2\pi^{-\frac{1}{2}}\int_0^\alpha \exp(-t^2)\,\mathrm{d}t.$$

For $\alpha = 0$ we have $f(\alpha) = 1-\pi/4$, and for $\alpha > 20$ $f(\alpha) \sim (4\alpha)^{-1}$. The function $f(\alpha)$ is tabulated for intermediate values of α by Robinson (1952). Its value is approximately $\frac{1}{2}(1-\pi/4)$ when $\alpha = \frac{3}{4}$.

The significance of Rack's calculation is that, in high-frequency vacuum tubes, the noise is determined primarily by the values of w_I and w_v at a point in the electron flow, near the cathode, where coherent interactions begin to exert an effect. We return to this topic in Chapter 15.

6 Noise and Quantum Mechanics

6.1. Introduction

ALTHOUGH in the majority of electronic systems the quantum $h\nu$, at the operating frequency ν, is much less than the thermal energy kT, so that the conventional Johnson-noise formula $\mathrm{d}V^2 = 4kTR\,\mathrm{d}\nu$ is adequate, this is not the case in masers and lasers. We therefore need formulae for noise which include quantum effects. This raises a number of difficult problems since, in both the Nyquist formula and its derivation, the notions of resistance dissipation, and available power play a central role, and these notions have no simple quantum-mechanical equivalents. The derivation given in Chapter 5, although (as we were at pains to point out) it applies to systems of quantum particles, does not contain Planck's constant in the final result. This is because, although the charge carriers were treated in general terms, the applied potential and its associated field were treated as classical variables. In fact, Nyquist in his original treatment did include the quantum statistical formula for the average energy of a oscillator, though not the zero-point energy term. If we add this term we get

$$\mathrm{d}V^2 = 4R\left\{\frac{h\nu}{\exp(h\nu/kT)-1}+\tfrac{1}{2}h\nu\right\}\mathrm{d}\nu, \tag{6.1}$$

and the available noise power in $\mathrm{d}\nu$ is

$$\mathrm{d}P = \left\{\frac{h\nu}{\exp(h\nu/kT)-1}+\tfrac{1}{2}h\nu\right\}\mathrm{d}\nu. \tag{6.2}$$

This is, however, unsatisfactory, for (6.2) is a consequence of (6.1), which includes the purely classical concept of a resistance R, and, even if we derive (6.2) directly, the notion of an available noise power implies a resistance in which this power can be dissipated. Therefore, we shall have to approach the problem *ab initio* and attempt to reformulate the theory in a way consistent with the basic concepts of quantum mechanics. This implies that we shall treat notions such as resistance and dissipation as explicitly statistical processes describing the coupling between one quantum system (the circuit) and another quantum system, (the heat-sink) which has so many modes that any energy transferred to it is irretrievably lost as disorganized energy. We shall achieve this end by considering the attenuation of signals and the generation of noise in a dissipative transmission line, and we shall phrase our discussion in such a form that it will be applicable later to systems with gain, such as a travelling-wave maser. The treatment is largely based on two papers by Gordon, Walker, and Louisell (1963) and Robinson (1965).

6.2. The uncertainty principle

In classical mechanics the motion of a set of particles interacting according to known dynamical laws is, at least in principle, determinate, although it must be admitted that in practice, except for the simplest systems, the exact details of the calculation have so far evaded us. Nevertheless, we can be reasonably certain that, if the coordinates and momenta of all the particles are known at $t = 0$, then the state of the system at a later time t is uniquely determined and that, if X is an observable quantity associated with the system, the time development of X from $X(0)$ to $X(t)$ is also uniquely determined. The fundamental difference in quantum mechanics is that, even in principle, we cannot specify the initial state of the system so that $X(0)$ has a definite value, and at the same time preserve determinate equations of motion for $X(t)$. This restriction is summarized in Heisenberg's uncertainty principle, and the laws of quantum mechanics form a scheme for obtaining statistical statements about $X(t)$ when $X(0)$ is given only with statistical precision. This is usually phrased as follows. If, at $t = 0$, the state of the system is specified with the utmost precision allowed by the uncertainty principle and (at $t = 0$) X and its powers have expectation values $\langle X(0)\rangle$, $\langle X^2(0)\rangle$, etc., then we can construct equations which yield the expectation values at later times.

In the Schrödinger picture we imagine the state of the system to be described by a function ψ which evolves according to Schrödinger's equation from $\psi(0)$ to $\psi(t)$. Definite rules then allow us to use ψ to calculate $\langle X\rangle$, $\langle X^2\rangle$, etc. at any instant. In the Heisenberg picture we fix our attention on the time development of these expectation values. The only state of the system that concerns us is the initial state, which specifies the starting conditions. Thus, in the Heisenberg picture, the expectation value of say $X(t)$ eventually will be expressed in terms of the expectation value of $X(0)$ and other variables, e.g. $Y(0)$ or $Z(0)$, associated with the initial state. A Heisenberg calculation, in fact, looks exactly like a classical calculation, although its interpretation is different.

The initial state $\psi(0)$ of the system is subject to the restrictions implied by the uncertainty principle. If the system consists of a single particle, and the observable quantities in question are its momentum components p_i ($i = 1, 2, 3$) and coordinates r_i, then the uncertainty principle states that no form for ψ can be found for which the quantity

$$\langle\Delta r_i^2\rangle\langle\Delta p_j^2\rangle < \left(\frac{h}{4\pi}\right)^2 \delta_{ij},$$

where $\langle \Delta r_i^2 \rangle = \langle r_i^2 \rangle - \langle r_i \rangle^2$ and $\langle \Delta p_j^2 \rangle = \langle p_j^2 \rangle - \langle p_j \rangle^2$ and δ_{ij} is the Kronecker δ. If we write $h/2\pi$ as $\hbar$ we can express this by saying that for any possible state

$$\langle \Delta r_i^2 \rangle \langle \Delta p_i^2 \rangle \geqslant (\hbar/2)^2. \tag{6.3}$$

This is usually stated loosely as implying that the product of the uncertainties in r_i and p_i is greater than or equal to $\hbar/2$. A state for which the equality holds is a minimum uncertainty state with respect to r_i and p_i.

In the language of classical mechanics, the quantities r_i and p_i are conjugate variables or canonically conjugate variables. We shall discuss what this means in the next section, but here we go on to generalize eqn (6.3). If p_i and q_i are two canonically conjugate variables (which may be mechanical or electrical variables) associated with a system, then, for any possible state of the system,

$$\langle \Delta p_i^2 \rangle \langle \Delta q_i^2 \rangle \geqslant (\hbar/2)^2. \tag{6.4}$$

We also note that, if p_i and q_i are regarded as linear operators acting on wavefunctions, eqn (6.4) is a consequence of the commutation relation

$$[q_i, p_i] \equiv q_i p_i - p_i q_i = \mathrm{i}\hbar, \tag{6.4}$$

where, on the right, $\mathrm{i} = (-1)^{\frac{1}{2}}$.

6.3. The Hamiltonian and Heisenberg equations of motion

The notion of pairs of canonically conjugate variables associated with a physical system arises in the Hamiltonian formulation of classical mechanics but we require a slightly more general definition which can be applied to electromagnetic systems.

The state of a classical system, e.g. a system of particles, can be specified in terms of a set of coordinates q_n ($n = 1, 2 \ldots N$) and their time derivatives $\dot{q}_n$. However, we could use the same description to deal with an electrical network if we took the q_n to be the charges on the capacitances in the system and the $\dot{q}_n$ to be currents. The equations of motion of the classical system can also be assumed to be known, i.e. they are Newton's laws or Kirchhoff's laws, and they take the form of a set of second-order differential equations. Using these equations as a guide we can construct a function $L(q_n, \dot{q}_n)$, known as the Lagrangian of the system, such that the equations of motion can be expressed as

$$\frac{\mathrm{d}}{\mathrm{d}t}\frac{\partial L}{\partial \dot{q}_n} - \frac{\partial L}{\partial q_n} = 0, \qquad n = 1, 2, \text{etc.} \tag{6.6}$$

For example, if a tuned circuit contains an inductance λ in parallel with a capacitance γ it is easy to see that $L = \frac{1}{2}\lambda\dot{q}^2 - q^2/2\gamma$, where q is the charge on γ. This leads to $\lambda\ddot{q} + q/\gamma = 0$, which is the usual circuit equation. For an isolated, conservative, mechanical system it is always possible to set $L = T - U$, where T is the kinetic energy and U the potential energy, and analogies drawn from this are a useful guide to a possible form for L. The acid test for the correctness of L is, however, that it should yield the correct equations of motion when substituted in (6.6). We now define the function H by

$$H = \sum_n \dot{q}_n \frac{\partial L}{\partial \dot{q}_n} - L, \tag{6.7}$$

and we have

$$\frac{\mathrm{d}H}{\mathrm{d}t} = \sum_n \left(\dot{q}_n \frac{\mathrm{d}}{\mathrm{d}t}\frac{\partial L}{\partial \dot{q}_n} + \ddot{q}_n \frac{\partial L}{\partial \dot{q}_n}\right) - \sum_n \left(\dot{q}_n \frac{\partial L}{\partial q_n} + \ddot{q}_n \frac{\partial L}{\partial \dot{q}_n}\right).$$

This, by virtue of eqn (6.6) is zero. Thus H is a constant of the motion. We also define the momentum canonically conjugate to q_n as

$$p_n = \partial L/\partial \dot{q}_n, \tag{6.8}$$

so that

$$H = \sum_n p_n \dot{q}_n - L. \tag{6.9}$$

If in this equation $\dot{q}_n$ is eliminated by being expressed in terms of the p_n and q_n, so that H becomes a function of the p_n and q_n alone, the function H is known as the Hamiltonian of the system. It is usually equal to the total energy function of the system, which is obviously a constant of the motion, but this is not necessary. From eqn (6.9) we have, for a small change in the p_n and q_n,

$$\mathrm{d}H = \sum_n \left(\dot{q}_n\,\mathrm{d}p_n + p_n\,\mathrm{d}\dot{q}_n - \frac{\partial L}{\partial \dot{q}_n}\mathrm{d}\dot{q}_n - \frac{\partial L}{\partial q_n}\mathrm{d}q_n\right).$$

The second and third terms cancel (in view of eqn (6.8)) and, in the last term, eqn (6.6) gives

$$\frac{\partial L}{\partial q_n} = \frac{\mathrm{d}}{\mathrm{d}t}\frac{\partial L}{\partial \dot{q}_n} = \dot{p}_n,$$

so that

$$\mathrm{d}H = \sum_n (\dot{q}_n\,\mathrm{d}p_n - \dot{p}_n\,\mathrm{d}q_n).$$

Since H is a function only of the p_n and q_n it follows that

$$\dot{q}_n = \partial H/\partial p_n \tag{6.10a}$$

and

$$\dot{p}_n = -\partial H/\partial q_n. \tag{6.10b}$$

These are Hamilton's equations of motion. They replace the N second-order differential equations (6.6) by $2N$ simultaneous first-order differential equations. For our tuned circuit example we have $p = \lambda\dot{q}$ and $H = \frac{1}{2}p^2/\lambda + \frac{1}{2}q^2/\gamma$, so that (6.10a) becomes $\dot{q} = p/\lambda$ and eqn (6.10b) becomes $\dot{p} = -q/\gamma$. It is not difficult to see that $-p$ is the voltage across the inductance λ.

In classical problems the use of Hamilton's formulation is rarely worthwhile in simple systems, but in complex systems the fact that the equations of motion can all be written down directly (using eqns (6.10a) and (6.10b)), once the Hamiltonian function has been found, often leads to a very considerable simplification. For our purposes, however, its most important feature is that it leads to a direct and general connection between the classical equations of a problem and the corresponding quantum equations. To formulate this connection we have first to define the Poisson bracket of two dynamical variables. If u and v are two such variables, their Poisson bracket is

$$(u, v) = \sum_n \left(\frac{\partial u}{\partial q_n}\frac{\partial v}{\partial p_n} - \frac{\partial u}{\partial p_n}\frac{\partial v}{\partial q_n}\right). \tag{6.11}$$

It is easy to verify by direct substitution that $(p_k, p_l) = 0$, $(q_k, q_l) = 0$, and

$$(q_k, p_l) = \delta_{kl}. \tag{6.12}$$

The total time derivative of a dynamical variable is

$$\frac{\mathrm{d}u}{\mathrm{d}t} = \frac{\partial u}{\partial t} + \sum_n \left(\frac{\partial u}{\partial q_n}\dot{q}_n + \frac{\partial u}{\partial p_n}\dot{p}_n\right)$$

and, if we use Hamilton's equations to eliminate $\dot{q}_n$ and $\dot{p}_n$, this gives

$$\frac{\mathrm{d}u}{\mathrm{d}t} = \frac{\partial u}{\partial t} + \sum_n \left(\frac{\partial u}{\partial q_n}\frac{\partial H}{\partial p_n} - \frac{\partial u}{\partial p_n}\frac{\partial H}{\partial q_n}\right) = \frac{\partial u}{\partial t} + (u, H). \tag{6.13}$$

Thus, if u is not an explicit function of the time t, its time derivative is simply its Poisson bracket with H and so

$$\frac{\mathrm{d}u}{\mathrm{d}t} = (u, H). \tag{6.14}$$

As a typical mechanical example we might let u be the z component of the total angular momentum of the system; in an electrical network it might be the voltage between two nodes or junctions in the network.

In quantum mechanics observable quantities correspond to linear operators, and if u is an observable quantity the same symbol is also used to denote its corresponding operator. The expectation value of u, when the system is in a state $|\psi\rangle$, is $\langle\psi|u|\psi\rangle$, where the mathematical procedure for evaluating this expression depends on whether u is represented as an operator acting on wavefunctions which are functions of coordinates in real space, in which case $\langle\psi|u|\psi\rangle = \int_{\mathrm{Vol}} \psi^* u \psi \, \mathrm{d}V$, or whether some more general representation is being used. Fortunately, as we shall see, this need not concern us. The connection between the quantum and classical equations is then (Dirac, 1958) obtained by replacing the Poisson bracket of two dynamical variables by their commutator $(uv - vu)$ according to the rule

$$[u, v] \equiv uv - vu = \mathrm{i}\hbar(u, v). \tag{6.15}$$

The equation of motion for the operator representing a dynamical variable is therefore

$$i\hbar\frac{\mathrm{d}u}{\mathrm{d}t} = uH - Hu. \tag{6.16}$$

Eqn (6.16) can also be written as

$$i\hbar\frac{\mathrm{d}u}{\mathrm{d}t} = [u, H]. \tag{6.17}$$

We also note that

$$[q_k, p_l] = i\hbar\,\delta_{kl}. \tag{6.18}$$

In the Heisenberg picture the state of the system is time-independent and the same as the initial state $|\psi(0)\rangle$; thus, if $u(t)$ is the operator for a dynamical variable u, its expectation value at t is

$$\langle u(t)\rangle = \langle\psi(0)|u(t)|\psi(0)\rangle. \tag{6.19}$$

Now if we consider a large enough group of dynamical variables and solve the Heisenberg equations of motion (6.17) for all these variables, we shall be able to express a typical observable operator, say $u(t)$, in terms of the operators $u(0)$, $v(0)$, $w(0)$, etc. at $t = 0$, thus obtaining

$$\langle u(t)\rangle = \langle\psi(0)|f\{u(0), v(0), \text{etc.}\}|\psi(0)\rangle = \langle f(u(0), v(0), \text{etc.})\rangle.$$

Thus we can relate the expectation values of observables at an arbitrary time t to the expectation values of some function of a set of observables at an earlier time $t = 0$. The wavefunctions, or states, $|\psi\rangle$ disappear from the problem completely. The remarkable result of the rule (6.15) and its consequences (6.16), (6.17), and (6.18) is that the resulting equations of motion for the operators representing observables are the classical equations of motion for the observables themselves. It then follows that the equations of motion of the expectation values of the quantum operators reproduce exactly the classical behaviour of the system. The one specifically quantum-mechanical result arises from the fact that, whereas in classical mechanics u and $\langle u\rangle$ are identical and $\langle u^2\rangle = u^2 = \langle u\rangle^2$ so that $\langle\Delta u^2\rangle = 0$, in quantum mechanics $\langle u^2\rangle \geqslant \langle u\rangle^2$ and $\langle\Delta u^2\rangle \geqslant 0$. It is this uncertainty, or spread in the possible values of u, that is significant in noise calculations.

We now give one simple example to illustrate these results, and consider a quantum-mechanical system, namely, a nucleus, with a fixed intrinsic angular momentum $\mathbf{l}$ and magnetic moment $\mathbf{m}$ in a steady magnetic field $\mathbf{B} = (0, 0, B)$ parallel to the z axis. It can be shown that $\mathbf{m}$ is necessarily parallel to $\mathbf{l}$, and so we set $\mathbf{m} = \gamma\mathbf{l}$. It can be shown further that $\mathbf{m}$ and $\mathbf{l}$ can be regarded as proper dynamical variables. The commutation rule for $\mathbf{l}$ obtained from (6.18) is most conveniently expressed as $\mathbf{l}\wedge\mathbf{l} = i\hbar\mathbf{l}$, and the Hamiltonian is $H = -\mathbf{m}.\mathbf{B} = -\gamma\mathbf{l}.\mathbf{B}$, or $H = -\gamma l_z B$. Thus the equations of motion for the components of $\mathbf{l}$ are

$$i\hbar\dot{l}_x = [l_x, H] = -\gamma B[l_x, l_z] = \gamma B(\mathbf{l} \wedge \mathbf{l})_y = i\hbar\gamma B l_y,$$
$$i\hbar\dot{l}_y = [l_y, H] = -\gamma B[l_y, l_z] = -\gamma B(\mathbf{l} \wedge \mathbf{l})_x = -i\hbar\gamma B l_x,$$
$$i\hbar\dot{l}_z = [l_z, H] = -\gamma B[l_z, l_z] = 0.$$

The factors $i\hbar$ cancel and leave the equations

$$\dot{l}_x = \gamma B l_y,$$
$$\dot{l}_y = -\gamma B l_x,$$
$$\dot{l}_z = 0.$$

These we recognize as identical with the components of the classical equation of motion $\dot{\mathbf{l}} = \mathbf{m} \wedge \mathbf{B} = \gamma\mathbf{l} \wedge \mathbf{B}$. If we set $\omega = \gamma B$, then the solution for either the operators, or the classical variables, is

$$l_x(t) = l_x(0)\cos\omega t + l_y(0)\sin\omega t,$$
$$l_y(t) = -l_x(0)\sin\omega t + l_y(0)\cos\omega t,$$
$$l_z(t) = l_z(0).$$

In the classical problem this is the final solution, but in the quantum problem we still have to write down expressions such as

$$\langle l_x(t)\rangle = \langle l_x(0)\rangle\cos\omega t + \langle l_y(0)\rangle\sin\omega t$$

and

$$\langle l_x^2(t)\rangle = \langle l_x^2(0)\rangle\cos^2\omega t + \langle l_y^2(0)\rangle\sin^2\omega t + \langle l_x(0)l_y(0) + l_y(0)l_x(0)\rangle\cos\omega t\sin\omega t.$$

Since there is no guarantee that $\langle l_x^2(t)\rangle = \langle l_x(t)\rangle^2$, the quantum solution gives essentially the probability distribution for $l_x(t)$ in terms of the probability distributions of $l_x(0)$ and $l_y(0)$, together with any cross-correlation terms.

Before we embark on any specific detailed calculations we now look in general terms at the implication of these results for noise in quantum-mechanical systems. Suppose that we have a system (e.g. some kind of amplifier) which produces a voltage output $V(t)$ of such a large amplitude that, at the output, all further noise processes, quantum-mechanical or otherwise, are irrelevant, but that somewhere, at an earlier stage in the system, there is a device requiring quantum treatment. The output $V(t)$ can be regarded as a dynamical variable related to this device. The final results of our calculation will be expressions for $\langle V(t)\rangle$, $\langle V^2(t)\rangle$, and $\langle\Delta V^2(t)\rangle$, and part at least of the uncertainty $\langle\Delta V^2(t)\rangle$ will arise from the limitations imposed by the laws of quantum mechanics on our calculations for the significant quantum-mechanical device. It is in this sense that fluctuations of quantum-mechanical origin have practical and experimental significance.

6.4. Dissipation in quantum mechanics

No serious problem is presented by the occurrence of friction or dissipation in classical systems. The transition from a discussion of the motion of a free particle under gravity in vacuum to a discussion of the same motion in a viscous medium can be made simply by the inclusion of additional terms in the equation of motion. We recognize, of course, that the mechanical energy of the particle is no longer conserved, but continually decreases as it is degraded to heat, and that the degraded energy plays no further role in the mechanical problem. We also recognize that in a closed dissipative system, with no external force-fields or sources of mechanical energy, all mechanical motion will eventually cease and the coordinates of all the particles take fixed values, their momenta taking zero values.

The situation is quite different in quantum mechanics. If we attempt to construct a dissipative equation of motion, the final state of the system eventually will violate the uncertainty principle. Dissipation is not a process, or concept, which plays any role in the microscopic quantum-mechanical description of a physical system. It is an entirely macroscopic concept, and its relation to quantum mechanics is a statistical relation. If we have, for example, a tuned circuit containing a resistance, the charge q on the capacitor is a perfectly legitimate quantity to use both as a dynamical variable and as one of a pair of canonically conjugate variables associated with the system. We might also regard q as the one interesting observable associated with the system. There are, however, many other pairs of conjugate variables associated with the system, amongst these we may mention the mechanical variables describing the charge carriers and, especially, the variables describing the state of excitation of the lattice modes of the resistive material in the circuit. Collisions couple the one interesting observable to these uninteresting and very numerous lattice variables. The system as a whole possesses a Hamiltonian which is constant in time, and this can be used to obtain the equation of motion for any variable of interest. The motion of the system will consist of a transfer of energy from mode to mode within the system. Energy transferred from the macroscopic, electric mode to the lattice modes is transferred from a single mode to upwards of 10^{20} modes. The probability that this energy will reappear in coherent form in the electric mode is nil in any finite time; the energy is dissipated. The 10^{20} lattice modes supply energy to the electric mode. The probability that this will result in a macroscopic oscillation which can be recognized as such, from the moment of its onset to its final decay, is also nil. This energy transfer is random and unpredictable and leads to noise.

Because the same fundamental microscopic collision processes mediate both the dissipation and the generation of noise, there is a direct connection between the two processes. In the classical limit this leads to Nyquist's Johnson-noise formulae.

We can extend these notions to include processes which, in a classical limit, would be called amplifier gain. In a quantum-mechanical amplifier we have to deal with three component parts of a complete quantum system. First, there is one mode or a small number of modes which describe the aspects (e.g. output voltage V) of interest to us. Secondly, there is a heat-sink, with a vast number of modes, which accounts for any dissipative process, and finally there is an active component with one, or possibly very many, modes excited from outside the system. The gain mechanism is provided by energy interchange between this active component, the signal modes, and the heat-sink. Because the interactions responsible for gain also transfer fluctuation between the component systems, the signal modes acquire fluctuations from the other two components, and part of the fluctuations in the output $V(t)$ derive from this source.

6.5. Noise in a lossy transmission line

We consider a long, otherwise-lossless line, coupled uniformly along its length to a heat-sink which provides an explicit description of the attenuation. A voltage generator $V(t)$ is located at the centre of the line and, if the line is long enough and lossy enough, the effects of $V(t)$ at the ends of the line

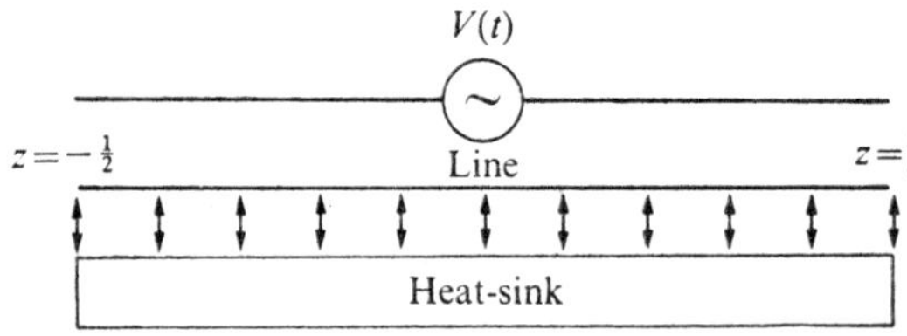

FIG. 6.1. Line uniformly coupled by a loss mechanism to a heat-sink.

will require quantum-mechanical treatment even if the input $V(t)$ itself is large. We can assume that this is the case and that $V(t)$ is so large that it can be treated as a classical parameter in the problem. The equations that we will have to use will be quite sufficiently complicated to make it worthwhile choosing a system of units in which (1) Planck's constant $\hbar = 1$, (2) the length of the line is unity and the line extends from $z = -\frac{1}{2}$ to $z = \frac{1}{2}$, and (3) the capacitance and inductance per unit length of the line are also unity. The system is illustrated in Fig. 6.1.

Now we must construct a Hamiltonian for this system, containing three terms: H_l pertaining to the lossless line, H_s pertaining to the heat-sink, and H_{ls} describing their interaction. In this Hamiltonian $V(t)$ will appear as a scalar parameter, but the voltage on the line will be a dynamical variable or operator. Because the system is continuous the line will have an infinite number of degrees of freedom, and we shall have to describe it in terms of a

Hamiltonian density $\mathscr{H}$ and Lagrangian density $\mathscr{L}$, so that

$$H_L = \int_{-\frac{1}{2}}^{\frac{1}{2}} \mathscr{H}\,dz \tag{6.20a}$$

and

$$L = \int_{-\frac{1}{2}}^{\frac{1}{2}} \mathscr{L}\,dz \tag{6.20b}$$

In a continuous system Lagrange's equation of motion becomes (Goldstein 1950).

$$\frac{d}{dt}\frac{\partial\mathscr{L}}{\partial\dot{q}}+\frac{d}{dz}\frac{\partial\mathscr{L}}{\partial q'}-\frac{\partial\mathscr{L}}{\partial q}=0,$$

where $q' = \partial q/\partial z$. We now take the coordinate q to be the total charge to the right of z, so that $\dot{q} = I$ is the current in the line at z and $q' = -V$ is the voltage across the line. We then automatically obtain $\partial I/\partial z = -\partial V/\partial t$, which is one of the transmission line equations. A Lagrangian density which yields the other is

$$\mathscr{L} = \tfrac{1}{2}(\dot{q}^2-q'^2)$$

for then, since $\partial\mathscr{L}/\partial q = 0$, Lagrange's equation becomes $\ddot{q}-q'' = 0$ or $\partial I/\partial t = -\partial V/\partial z$. Having thus found the correct form for L we have, for the canonical variable p conjugate to q, $p = \partial\mathscr{L}/\partial\dot{q} = \dot{q}$ and so

$$\mathscr{H} = p\dot{q}-\mathscr{L} = \tfrac{1}{2}(p^2+q'^2).$$

If we divide the line into cells of infinitesimal length, at positions z_n, then the variables $q(n) = q(z_n)$, for different n, form a set of independent coordinates for the electrical problem and the fundamental commutator is

$$[q(n), p(m)] = \mathrm{i}\,\delta_{nm}.$$

In the continuous limit this is replaced by

$$[q(z), p(z')] = \mathrm{i}\,\delta(z-z'), \tag{6.21}$$

where $\delta(z-z')$ is the Dirac delta function. We next expand $p(z, t)$ and $q(z, t)$ in terms of a complete set of orthonormal running waves so that, for example,

$$p(z, t) = \sum_{k=0}^{\infty}(P_k(t)u_k(z)+P_k^{+}(t)u_k^{*}(z)), \tag{6.22}$$

where

$$u_k(z) = \exp(\mathrm{i}\omega_k z) = \exp(2\pi \mathrm{i}kz) \tag{6.23}$$

is an ordinary algebraic function, and the coefficients $P_k(t)$, etc. are operators representing the dynamical variables. We have, for example,

$$P_k(t) = \int_{-\frac{1}{2}}^{\frac{1}{2}} p(z, t)\exp(-2\pi ikz)\,dz, \qquad k > 0,$$

and it is easy to verify that all the coefficients P_k and Q_k commute except for the two pairs

$$[Q_k^+, P_k] = \mathrm{i} \tag{6.24a}$$

and

$$[Q_k, P_k^+] = \mathrm{i}. \tag{6.24b}$$

We now introduce the operators

$$A_k(t) = \left(\frac{1}{2\omega_k}\right)^{\frac{1}{2}}(P_k(t) - \mathrm{i}\omega_k Q_k(t)), \qquad A_k^+(t) = \left(\frac{1}{2\omega_k}\right)^{\frac{1}{2}}(P_k^+(t) + \mathrm{i}\omega_k Q_k^+(t)) \tag{6.25a}$$

and

$$B_k(t) = \left(\frac{1}{2\omega_k}\right)^{\frac{1}{2}}(P_k^+(t) - \mathrm{i}\omega_k Q_k^+(t)), \qquad B_k^+(t) = \left(\frac{1}{2\omega_k}\right)^{\frac{1}{2}}(P_k(t) + \mathrm{i}\omega_k Q_k(t)). \tag{6.25b}$$

Their only non-vanishing commutators are

$$[A_k(t), A_k^+(t)] = [B_k(t), B_k^+(t)] = 1. \tag{6.26}$$

The Hamiltonian density for the line can be expressed in terms of these operators as

$$H_l = \int_{-\frac{1}{2}}^{\frac{1}{2}} \mathscr{H}_l \, \mathrm{d}z = \sum_k \omega_k \{A_k^+(t)A_k(t) + B_k^+(t)B_k(t) + 1\}, \tag{6.27}$$

while the operators corresponding to current and voltage are

$$I(z, t) = \sum_k \left(\frac{\omega_k}{2}\right)^{\frac{1}{2}} \{(A_k(t) + B_k^+(t))u_k(z) + \text{H.c.}\}, \tag{6.28}$$

and

$$V(z, t) = \sum_k \left(\frac{\omega_k}{2}\right)^{\frac{1}{2}} \{(A_k(t) - B_k^+(t))u_k(z) + \text{H.c.},\} \tag{6.29}$$

where H.c. denotes the Hermitian conjugate.

The time derivatives of the operators $A_k(t)$, $A_k^+(t)$, $B_k(t)$, and $B_k^+(t)$ can be obtained from $\mathrm{i}\dot{A}_k(t) = [A_k(t), H]$, etc. and, with the aid of the commutation rules (6.26) and the result (6.27), we obtain

$$\dot{A}_k(t) = -\mathrm{i}\omega_k A_k(t),$$

$$\dot{B}_k(t) = -\mathrm{i}\omega_k B_k(t),$$

$$\dot{A}_k^+(t) = \mathrm{i}\omega_k A_k^+(t),$$

and

$$\dot{B}_k^+(t) = \mathrm{i}\omega_k B_k(t).$$

Since, in I and V, the operators $A_k(t)$ are associated with $u_k(z)$ for which $\partial u_k/\partial z = \mathrm{i}\omega_k u_k$, while $A_k^+(t)$ is associated with $u_k^*(z)$ for which

$$\partial u_k^*/\partial z = -\mathrm{i}\omega_k u_k^*,$$

the contributions to I and V arising from terms in $A_k(t)$ and $A_k^+(t)$ both satisfy $\partial I/\partial t = -\partial I/\partial z$ and so correspond to forward waves. Conversely, $B_k(t)$ is associated with $u_k^*(z)$ and $B_k^+(t)$ with $u_k(z)$, so that these terms correspond to backward waves.

In the Schrödinger picture the operators A_k^+ and B_k^+ would be creation operators for forward and backward waves, and A_k and B_k would be the corresponding annihilation operators. In the Heisenberg picture they are essentially the operators corresponding to the classical amplitudes of these waves in the sense that $A_k^+A_k$ and $B_k^+B_x$ correspond to their intensities.

The heat-sink is a complicated dynamical system, but we can assume that it is equivalent to the lattice vibrators of a solid and can be described as a sum of harmonic-oscillator terms. Eventually we shall regard these as a continuous distribution of oscillators. If we let $S_{pq}(t)$ be the Heisenberg time-dependent operator equivalent to the Schrödinger annihilation operator for an oscillator of frequency ω_q located at $z = p$, the heat-sink Hamiltonian can be expressed as

$$H_s = \sum_p \sum_q \omega_q(S_{pq}^+(t)S_{pq}(t)+\tfrac{1}{2}), \tag{6.30}$$

and the operators S_{pq} and S_{pq}^+ satisfy the commutation rule.

$$[S_{pq}(t), S_{pq}^+(t)] = 1. \tag{6.31}$$

A heat-sink oscillator of frequency ω_q at $z = p$ couples to the wave amplitude at $z = p$ and, if we restrict our attention to terms of first order in the coupling constant and note that any interaction Hamiltonian must be Hermitian, we find that the most general form we need consider is

$$H_{ls} = \sum_k \sum_p \sum_q \alpha_{kq}[\{A_k(t)S_{pq}^+(t)+B_k^+(t)S_{pq}(t)\}u_k(p)+\text{H.c.},] \tag{6.32}$$

where, because the coupling is uniform, α_{kq} the coupling constant does not depend on the position p. Furthermore the α_{kq} are small (compared with ω_q) real constants.

We now introduce the classical signal source $V(t)$ by an additional term H_d in $H = H_l+H_s+H_{ls}+H_d$. If $V(t)$ is first expressed as

$$V(t) = V_0 \exp(-i\omega t)+V_0^* \exp(i\omega t) \tag{6.33a}$$

and we note that it delivers power to the line at a rate $V(t)\,I(0, t)$, we obtain

$$V(t)I(0, t) = \frac{dH}{dt} = \frac{\partial H}{\partial t} = \frac{\partial H_d}{\partial t}. \tag{6.33b}$$

If $I(0, t)$ is expressed in terms of the operators A_k etc. and we remember that the time-dependence of H on t, through the time variation of these operators, is not included in $\partial H/\partial t$, we can integrate (6.33b) immediately, to obtain

$$H_d = F(t)I(0, t) = F(t) \sum_k \left(\frac{\omega_k}{2}\right)^{\frac{1}{2}}\{A_k(t)+B_k^+(t)+\text{H.c.}\}, \tag{6.34}$$

where $F(t)$ is the ordinary algebraic function

$$F(t) = \frac{\mathrm{i}}{\omega}\{V_0 \exp(-\mathrm{i}\omega t) - V_0^* \exp(\mathrm{i}\omega t)\}. \tag{6.35}$$

The Heisenberg equation of motion for $A_k(t)$ is

$$\mathrm{i}\dot{A}_k(t) = [A_k(t), H],$$

and if we now assemble all the terms in H, this gives

$$\dot{A}_k(t) + \mathrm{i}\omega_k A_k(t) = -\mathrm{i}\left(\frac{\omega_k}{2}\right)^{\frac{1}{2}} F(t) - \mathrm{i}\sum_p\sum_q \alpha_{kq} S_{pq}(t) u_k^*(p). \tag{6.36}$$

We take the Laplace transform of this equation, with the definitions

$$\bar{A}_k = \int_0^\infty A_k(t) \exp(-\theta t)\, dt,$$

$$A_k(t) = \frac{1}{2\pi \mathrm{i}} \int_{a-\mathrm{i}\infty}^{a+\mathrm{i}\infty} \bar{A}_k \exp(\theta t)\, \mathrm{d}\theta,$$

and let the letters A_k, etc. without bars denote values of operators at $t = 0$. We than have

$$(\theta + \mathrm{i}\omega_k)\bar{A}_k = A_k - \mathrm{i}\left(\frac{\omega_k}{2}\right)^{\frac{1}{2}} \bar{F} - \mathrm{i}\sum_p\sum_q \alpha_{kq} \bar{S}_{pq} u_k^*(p). \tag{6.37a}$$

The corresponding equations for the other variables are

$$(\theta + \mathrm{i}\omega_k)\bar{B}_k = B_k - \mathrm{i}\left(\frac{\omega_k}{2}\right)^{\frac{1}{2}} \bar{F} - \mathrm{i}\sum_p\sum_q \alpha_{kq} \bar{S}_{pq} u_k(p) \tag{6.37b}$$

and

$$(\theta + \mathrm{i}\omega_q)\bar{S}_{pq} = S_{pq} - \mathrm{i}\sum_j \alpha_{jq}\{\bar{A}_j u_j(p) + \bar{B}_j u_j^*(p)\}. \tag{6.37c}$$

We can use (6.37c) to eliminate $\bar{S}_{pq}$ from (6.37a) and, with the aid of the orthogonality relations $\sum_p u_k(p) u_j^*(p) = \delta_{kj}$, etc. we obtain

$$\left(\theta + \mathrm{i}\omega_k + \sum_q \frac{\alpha_{kq}^2}{\theta + \mathrm{i}\omega_q}\right)\bar{A}_k = A_k - \mathrm{i}\left(\frac{\omega_k}{2}\right)^{\frac{1}{2}} \bar{F} - \mathrm{i}\sum_p\sum_q \frac{\alpha_{kq} S_{pq} u_k^*(p)}{\theta + \mathrm{i}\omega_q}. \tag{6.38}$$

On the left the quantity $\sum \alpha_{kq}^2/(\theta + \mathrm{i}\omega_q)$ has zeros and poles at intervals along the imaginary frequency axis and causes $\bar{A}_k$ to have poles and zeros along this axis. The fact that these singularities lie on the imaginary axis (instead of to the left of it) means that the energy of excitation of the line remains forever within the system, a result built into the present calculation from the beginning, since energy can be exchanged between the line and the heat-sink oscillators but can never leave the entire system. The closeness of the spacing of the heat-sink oscillator frequencies ω_q, however, defines a Poincaré, or recurrence, cycle for the system, and in a real system the length of this cycle is all but infinite. If the system is examined over finite times, energy initially given to the line becomes randomized amongst many heat-sink oscillators and, from the point of view of the line, this energy is lost. Thus, for finite times, the net effect of coupling the line to the oscillators is to produce a decay of the excitation of the line which approximates to a simple exponential loss. This behaviour can be obtained mathematically from (6.38), but the proof is not trivial. The proof may be found in Robinson (1965), where it is shown that, as the distribution of the heat-sink frequencies approaches a continuum, the sum may be replaced by

$$\sum_q \frac{\alpha_{pq}^2}{\theta+\mathrm{i}\omega_q} \to \pi\alpha_k^2\sigma(\omega_k) = \mu_k, \tag{6.39}$$

where α_k represents α_{kk} and $\sigma(\omega_k)$ is the spectral density of heat-sink oscillators near ω_k. We note that μ_k is real and positive. We now have

$$(\theta+\mathrm{i}\omega_k+\mu_k)\bar{A}_k = A_k - \mathrm{i}\left(\frac{\omega^k}{2}\right)^{\frac{1}{2}}\bar{F} - \mathrm{i}\sum_p\sum_q \frac{\alpha_{kq}S_{pq}u_k^*(p)}{\theta+\mathrm{i}\omega_q}, \tag{6.40}$$

and there are similar equations for $\bar{B}_k$, $\bar{A}_k^+$, and $\bar{B}_k^+$.

The term $\bar{F}$ which arises from the signal $V(t)$ can be shown, using (6.35) and the definition of the Laplace transform, to be

$$\bar{F} = \frac{\mathrm{i}}{\omega}\left(\frac{V_0}{\theta+\mathrm{i}\omega} - \frac{V_0^*}{\theta-\mathrm{i}\omega}\right), \tag{6.41}$$

and the contribution it makes to $A_k(t)$ is obtained by inverting the Laplace transform

$$\bar{F}_k = -\frac{\mathrm{i}\left(\frac{\omega_k}{2}\right)^{\frac{1}{2}}\bar{F}}{\theta+\mathrm{i}\omega_k+\mu_k}. \tag{6.42}$$

The contribution this makes to the signal voltage at z is

$$V_F(z,t) = \sum_k \left(\frac{\omega_k}{2}\right)^{\frac{1}{2}} F_k(t)u_k(z), \tag{6.43}$$

and it is easy to verify that this contains no contribution from V_0^* in (6.41). The term in V_0 yields

$$F_k(t) = \left(\frac{\omega_k}{2}\right)^{\frac{1}{2}} \frac{V_0}{\omega}\left[\frac{\exp\{-\mathrm{i}\omega t\} - \exp\{-(\mathrm{i}\omega_k+\mu_k t)t\}}{\mathrm{i}(\omega_k-\omega)+\mu_k}\right]. \tag{6.44}$$

This is to be inserted in (6.43) and, if the sum is evaluated as an integral, so that

$$V_F(z,t) = \tfrac{1}{2}V_0 \int \frac{\omega_k[\exp\{-\mathrm{i}\omega t\} - \exp\{-(\mathrm{i}\omega_k+\mu_k)t\}]}{\mathrm{i}(\omega_k-\omega)+\mu_k} \exp\{\mathrm{i}\omega_k z\} \frac{\mathrm{d}\omega_k}{2\pi},$$

then, when we use the calculus of residues to evaluate the integral, we note that the pole is above the real axis. Thus when $z < 0$, so that we have to close the contour below the real axis, the result is zero. When $z > t$ and the contour is closed above the real axis the residues due to the two terms in the integral cancel, and the result again is zero. However, when $0 < z < t$ one term gives zero and the other gives the one non-vanishing contribution

$$V_F(z,t) = \tfrac{1}{2}V_0 \exp\{-\mathrm{i}\omega(t-z)-\mu z\}, \qquad 0 < z < t, \tag{6.45}$$

where $\mu = \mu_k$ at $\omega_k = \omega$. This is of course the classical causal result for a line whose attenuation constant is μ.

The remaining terms in (6.40) lead to noise, and, if we now omit the signal term, the Laplace inversion is

$$A_k(t) = A_k \exp\{-(\mathrm{i}\omega_k+\mu_k)t\} - $$
$$-\mathrm{i}\sum_p \sum_q \frac{\alpha_{kq} S_{pq} u_k^*(p)[\exp\{-\mathrm{i}\omega_q t\} - \exp\{-(\mathrm{i}\omega_k+\mu_k)t\}]}{\mathrm{i}(\omega_k-\omega_q)+\mu_k}. \tag{6.46}$$

Since the expectation values of the operators A_k and S_{pq} on the right are zero, the expectation value of $A_k(t)$, and therefore of the forward wave on the line for $z > 0$, is also zero. On the other hand, we have

$$\langle V_F^2(z,t)\rangle = \tfrac{1}{2}\sum_k \sum_j (\omega_k\omega_j)^{\frac{1}{2}} \langle A_k(t)A_j^+(t)\rangle u_k(z)u_j^*(z) + \text{H.c.}, \tag{6.47}$$

and this does not vanish.

The term in $\langle V_F^2\rangle$, which arises from A_k in (6.46) and which depends on the initial state of the line, is

$$\langle V_{AF}^2(z,t)\rangle = \sum_k \tfrac{1}{2}\omega_k \exp(-2\mu_k t)(\langle A_k A_k^+\rangle + \langle A_k^+ A_k\rangle). \tag{6.48}$$

We see that eventually all memory of the initial state of the line is obliterated. The term which involves the initial state of the heat-sink is

$$\langle V_{Fs}^2(z,t)\rangle = \frac{1}{2}\sum_p \sum_q \langle S_{pq}^+ S_{pq}\rangle \sum_k \sum_j (\omega_k\omega_j)^{\frac{1}{2}} \alpha_{kq}\alpha_{jq} \times$$
$$\times \frac{[\exp\{-\mathrm{i}\omega_q t\} - \exp\{-(\mathrm{i}\omega_k+\mu_k)t\}]u_k(z-p)}{\mathrm{i}(\omega_k-\omega_q)+\mu_k} \times$$
$$\times \frac{[\exp\{\mathrm{i}\omega_q t\} - \exp\{(\mathrm{i}\omega_j-\mu_j)t\}u_j^*(z-p)}{\mathrm{i}(\omega_q-\omega_j)+\mu_j} + \text{H.c.} \tag{6.49}$$

When the sums over the line oscillator frequencies k and j are evaluated as contour integrals, the results for each term, labelled by p, is zero unless $z-t < p < z$. This is clearly consistent with causality. An oscillator at p only affects the voltage at z if its effects can propagate to z in a time t. The final result is

$$\langle V_{F\mathrm{s}}^2(z,t)\rangle = \tfrac{1}{2}\sum_{p=z-t}^{z}\sum_{q}\langle S_{pq}S_{pq}^{+}+S_{pq}^{+}S_{pq}\rangle\omega_q\alpha_q^2\exp\{-2\mu_q(z-p)\}. \quad (6.50)$$

If the heat-sink is in thermal equilibrium at a uniform temperature T_{s} the expectation value in (6.50) is

$$\langle S_{pq}S_{pq}^{+}+S_{pq}^{+}S_{pq}\rangle = 1+\frac{2}{\exp(\omega_q/T_{\mathrm{s}})-1}, \quad (6.51)$$

and, when (6.50) is summed over the allowed range of p, i.e. integrated over the range $p = z-t$ to $p = z$, this leads to

$$\langle V_{F\mathrm{s}}^2(z,t)\rangle = \frac{1}{2}\sum_{q}\omega_q\frac{\alpha_q^2}{2\mu_q}\left\{1+\frac{2}{\exp(\omega_q/T_{\mathrm{s}})-1}\right\}\{1-\exp(-2\mu_q t)\}. \quad (6.52)$$

Similarly, if the line is initially at a temperature T_1, eqn (6.48) yields

$$\langle V_{FA}^2(z,t)\rangle = \frac{1}{2}\sum_{k}\omega_k\left\{1+\frac{2}{\exp(\omega_k/T_l)-1}\right\}\exp(-2\mu_k t). \quad (6.53)$$

Now the spectral density of heat-sink oscillators near $\omega_q = \omega$ is

$$\sigma(\omega) = \mu_q/\pi\alpha_q^2,$$

and so the contribution to $V_{F\mathrm{s}}^2(z,t)$ from frequencies in a range $\mathrm{d}\omega$ about ω is

$$\mathrm{d}V_{F\mathrm{s}}^2(z,t) = \tfrac{1}{2}\omega\left\{1+\frac{2}{\exp(\omega/T_{\mathrm{s}})-1)}\right\}\{1-\exp(-2\mu t)\}\frac{\mathrm{d}\omega}{2\pi},$$

and the similar contribution from (6.53) is

$$\mathrm{d}V_{FA}^2(z,t) = \tfrac{1}{2}\omega\left\{1+\frac{2}{\exp(\omega/T_l)-1)}\right\}\exp(-2\mu t)\frac{\mathrm{d}\omega}{2\pi}.$$

The sum of these terms contains a time-independent contribution

$$\mathrm{d}V_{F0}^2 = \tfrac{1}{2}\omega\,\mathrm{d}\omega/2\pi \quad (6.54)$$

which is the same at all points in the line whatever the values of T_l and T_{s}. It represents the irreducible uncertainty in V_F required by the uncertainty principle. We also note that, as the contribution to this term arising from the quantum-mechanical uncertainty in the initial specification of the state of the line decays, it is replenished by a term derived from the same quantum uncertainty in the heat-sink. In the next section we shall discuss the practical significance of this result.

Whatever it is initially, the contribution to V_F^2 from the initial state of the line eventually decays to zero and, since (6.54) is the only result relating to transient conditions that we shall require, we now go on to consider steady-state conditions, as $t \to \infty$, in which case the total fluctuations are to be derived from the heat-sink contribution (6.50) which becomes

$$\langle V_F^2(z, t)\rangle = \sum_{p=-\frac{1}{2}}^{z} \sum_{q} \langle S_{pq}S_{pq}^{+} + S_{pq}^{+}S_{pq}\rangle \omega_q \alpha_q^2 \exp\{-2\mu_q(z-p)\}. \quad (6.55)$$

We now come to the main point of this entire calculation. The system shown in Fig. 6.1 can be regarded as a signal source $V(t)$, of internal impedance equal to the characteristic impedance of the line and represented by the line to the left of $V(t)$. This is connected to a matched load represented by the line to the right of $V(t)$. Thus we have found a system which gives a description, compatible with quantum mechanics, of a signal generator with a real (dissipative) internal impedance. Let us now suppose that the heat-sink to the left of $V(t)$ at $z < 0$ is at a temperature T_1, and the sink to the right of $V(t)$ is at T_2 then, when we evaluate (6.55) using (6.51) and integrating over p, we obtain

$$\langle V_F^2(z)\rangle = \sum_q \left[\tfrac{1}{2}\omega_q + \frac{\omega_q \exp(-2\mu_q z)}{\exp(\omega_q/T_1)-1} + \frac{\omega_q\{1-\exp(-2\mu_q z)\}}{\exp(\omega_q/T_2)-1}\right]\frac{\alpha_q^2}{2\mu_q}, \quad (6.56)$$

where, in the second term, we have replaced a factor $1-\exp(-\mu_q)$ by unity, on the grounds that μ_q, the attenuation for the whole of the line to the left of the generator $V(t)$, is large. In (6.56) the first term represents the intrinsic quantum-mechanical fluctuations, the second term represents thermal noise generated in the internal impedance of the source and attenuated by $\exp(-2\mu_q z)$ by the time it reaches z, and the third term represents thermal noise generated to the right of $V(t)$. Because $\alpha_q^2/2\mu_q$ is related to the spectral density of the heat-sink oscillators we can go straight from (6.56) to the spectral density or power spectrum of V_f. If, in doing this, we introduce practical units, replace ω by $\nu = \omega/2\pi$, and let the characteristic impedance of the line be Z_0, we obtain $\mathrm{d}V_f^2(z) = w(\nu)\,\mathrm{d}\nu$, where the power spectrum is

$$w(\nu) = \tfrac{1}{2}h\nu Z_0 + \frac{h\nu Z_0 \exp(-2\mu z)}{\exp(h\nu/kT_1)-1} + \frac{h\nu Z_0\{1-\exp(-2\mu z)\}}{\exp(h\nu/kT_2)-1}. \quad (6.57)$$

We discuss the interpretation of this result in the next section, but we note that $w(\nu)$ is the power spectrum of the voltage fluctuations in the forward wave at z, not of the open-circuit voltage fluctuations. In the classical limit, with μz large, eqn (6.57) reduces to kT_2Z_0 and not $4kT_2Z_0$.

6.6. Quantum fluctuations and circuit concepts

Eqn (6.45) of the last section shows that, if a large coherent signal, describable in classical terms, is injected into an attenuating system, then even at a point where it has decreased to a level comparable with quantum fluctuations it still behaves classically. The expectation value of the voltage at this point is the value we should derive classically from the input and the attenuation of the system. Thus, for example, if a radio transmitter generates a large output, the much weaker output at the terminals of the aerial of a distant receiver can be treated classically. In general, signals behave classically whatever their magnitude.

Eqn (6.57) is the other relevant result, and it has two aspects. First consider the fluctuations at $z = 0$. These correspond to the fluctuations at the output terminals of a signal generator of internal impedance Z_0 at a temperature T when it is matched to a load Z_0 but, being derived from the forward or outward wave alone, they are not the entire fluctuations at this point. They omit fluctuations generated in the load. Nevertheless, in the language of circuit theory we should say that the available noise power in $d\nu$ was

$$dP = \frac{dV^2}{Z_0} = \left\{\tfrac{1}{2}h\nu + \frac{h\nu}{\exp(h\nu/kT_1)-1}\right\} d\nu. \tag{6.58}$$

If we had a perfect amplifier, not itself subject to quantum-mechanical restrictions, this would be the signal power required for unity signal to noise ratio. As we shall see in Chapter 17, a maser amplifier, which is the nearest we can come to a perfect amplifier but which is, like all real amplifiers, subject to quantum-mechanical laws, leads to a different result for the smallest detectable signal. Nevertheless, eqn (6.58) interpreted in conventional circuit terms correctly describes the part of the fluctuations due to the source.

The second aspect of (6.57) relates to a system such as a distant (classical) powerful radio transmitter and the signal and noise available at the aerial terminals of the receiver. In this case if the signal has been transmitted through a lossy medium at T_2, the noise from the generator will have disappeared and the available noise power is

$$dP = dV^2/Z_0 = \left\{\tfrac{1}{2}h\nu + \frac{h\nu}{\exp(h\nu/kT_2)-1}\right\} d\nu. \tag{6.59}$$

In the same system the signal is, of course, classical throughout.

It would appear then that the quantum-mechanical generalization of Nyquist's classical result $dP = kT\,d\nu$ is just (6.58) or (6.59). This is not quite true. If the temperature T_2 is high, the quantum-mechanical features of (6.59) are irrelevant and the noise power is very nearly $kT_2\,d\nu$; if, however, T_2 is low and the finite value of h is significant, we shall only detect its effects if we use an amplifier so refined that it is subject to quantum-mechanical restrictions itself. Thus, simply on the basis of (6.59), we cannot state that the minimum signal power detectable with an amplifier of bandwidth $d\nu$ is given by (6.59) and that, as $T_2 \to 0$, this power reduces the $\tfrac{1}{2}h\nu\,d\nu$. We have first to investigate whether amplification itself introduces noise. This can be done by an extension of the treatment in the last section (Robinson, 1965), and it can be deduced also that the minimum signal is $h\nu\,\mathrm{d}\nu$, by a general argument based on the uncertainty principle (Heffner, 1962). We shall deduce the same result, using (6.59), in Chapter 17). With this restriction eqn (6.59) is an adequate basis for the discussion of noise in quantum-mechanical systems.

Although the term $\frac{1}{2}h\nu\,\mathrm{d}\nu$ in (6.59) looks like transferable, or available, noise power, we must remember that for this to have any meaning it must be transferable to a real physical system obeying the laws of quantum mechanics. This system will itself have fluctuations of the same nature. As we have seen (6.52, 6.53), as fast as these fluctuations are transferred from one part of a system to a second part, they are replenished by fluctuations transferred in the opposite direction. Regarded as a statement about the voltage fluctuations in a single system the equation

$$\mathrm{d}V^2 = (\tfrac{1}{2}Z_0 h\nu + \text{etc.})\,\mathrm{d}\nu$$

is unexceptionable but, regarded as a statement about the *observable* fluctuations, it is insufficient. Here we make contact with a fundamental quantum principle. Statements about physical quantities are incomplete without a discussion of the means by which they are to be observed.

7 Flicker Noise

AT frequencies above about 10^5 Hz the noise associated with most electronic devices has a white power spectrum and can be adequately described in terms of thermal or shot-noise processes within the device. The noise at low frequencies, however, often considerably exceeds the value expected from these processes and the power spectrum increases as the frequency decreases. Generally, this excess noise is known as flicker noise, and in many cases its power spectrum is inversely proportional to the frequency. The frequency at which the excess noise is equal to the normal white noise varies very considerably from device to device, in some bipolar transistors under optimum bias conditions it may be as low as 1 Hz, whereas in microwave point-contact diodes it may be as high as 100 MHz. Usually it is in the region of a few kilohertz. The magnitude of the excess noise is also very variable, and two devices of nominally the same specification may differ by factors of 3 or more. This suggests that flicker noise is often associated with fine details of the device structure that are not under the manufacturer's control. In many cases it is almost certainly a surface effect in devices whose signal response is controlled by volume effects. This has a rather curious consequence for, since very high-frequency devices call for a stricter control of device parameters, the best low-frequency devices are often those intended for very high-frequency use. This is the case with both vacuum triodes and junction field-effect transistors. In triodes, flicker noise originates in gross fluctuations of the cathode emission, and in microwave triodes, where the cathode-to-grid spacing may be less than 10 μm, the cathode has to be very carefully and uniformly prepared. In field-effect transistors, where drain-to-gate leakage contributes to the flicker noise, control of conditions in this region, which is necessary to minimize the reverse Miller capacitance, results in reduced flicker noise. This state of affairs is less common in bipolar transistors, where good high-frequency response requires a base structure with a large ratio of periphery to area (to minimize the base resistance), and this militates against careful control of the base region and its surface condition.

Any device in thermal equilibrium is directly subject to the laws of thermodynamics, and so its noise output is solely Johnson noise with a white spectrum. Thus flicker noise can only occur in non-equilibrium situations in devices subjected to applied bias voltages or bias voltages derived from a signal.

If a device generates noise by a process which, though random, is characterized by a relaxation time τ_n, the correlation function for the noise will be of the form $A_n \exp(-t/\tau_n)$, where A_n is constant. This leads to a power spectrum

$$w(f) = 4\int_0^\infty A_n \exp(-t/\tau_n)\cos 2\pi ft\, dt = \frac{4A_n\tau_n}{1+(2\pi f\tau_n)^2}. \tag{7.1}$$

At low frequencies this gives $w(f)$ varying as $1/f^2$ rather than as $1/f$, and so flicker noise cannot be explained in terms of noise processes associated with any single relaxation time. A distribution of relaxation times such that the normalized probability of a time between τ and $\tau+d\tau$ is $g(\tau)\,d\tau$ yields

$$w(f) = \int_0^\infty \frac{4A\tau g(\tau)\,d\tau}{1+(2\pi f\tau)^2}. \tag{7.2}$$

van der Ziel (1959) considers the case where τ is controlled by an activation energy E so that τ and E are related by

$$\tau = \tau_0 \exp(E/kT), \tag{7.3}$$

where τ_0 is a constant. If all energies between E_1 corresponding to τ_1 and E_2 corresponding to τ_2 are equally likely then, since

$$E = kT\ln \tau/\tau_0 \quad \text{and} \quad dE = kT\,d\tau/\tau,$$

the normalized probability is

$$g(\tau)\,d\tau = \frac{d\tau}{\tau \ln \tau_2/\tau_1}, \tag{7.4}$$

and this yields

$$w(f) = \frac{A}{2\pi \ln \tau_2/\tau_1}\cdot\frac{\tan^{-1}2\pi f\tau_2 - \tan^{-1}2\pi f\tau_1}{f}. \tag{7.5}$$

Provided that $f\tau_2$ is large and $f\tau_1$ is small the power spectrum is inversely proportional to f over an appreciable range of frequencies. The result is not greatly altered if the distribution of activation energies is not completely uniform.

Since, in many cases, the observed power spectrum is proportional to $1/f$ down to very low frequencies of the order of 10^{-2} Hz or less, it is clear that the longest relaxation times τ_2 must be very long indeed.

The r.m.s. flicker noise in some components, such as the carbon microphone and the composition resistor, is found to be proportional to the d.c. bias current or voltage. It is then natural to attribute the noise to fluctuations in the contact resistance between granules. The noise performance of this type of circuit element can be expressed in terms of the ratio of the r.m.s. fluctuating voltage to the applied bias voltage (when this is derived from a constant-current or high-impedance source). In technical specifications this is sometimes presented for a bandwidth of 1 Hz at a specific frequency, but more usually in terms of the noise integrated over a band of frequencies, e.g. from 1 Hz to 1000 Hz. A typical figure for the integrated noise in resistors is 1 μV per volt.

If two similar devices displaying flicker noise and working under the same conditions are connected in parallel, the mean square noise current and the mean current are both doubled, so that the ratio of the r.m.s. noise to the mean current is reduced by $1/\sqrt{2}$. Sometimes this can be put to practical use, e.g. by using a number of triodes or transistors in parallel. It also suggests that, of two devices of generally similar construction, the one presenting the greater cross-sectional area for current flow will have the smaller noise, for the same current. This appears to be approximately true for vacuum tubes, but of doubtful relevance to semiconductor devices.

In vacuum tubes, flicker noise arises from slow random changes in the electron emissivity of the cathode surface, in composition resistors from fluctuations in the contact resistance between granules, and in some types of ceramic capacitor from thermally activated leakage currents through the bulk of the material. In biased semiconductor devices, it is associated mainly with the effects of generation and recombination processes which either produce or eliminate minority carriers. These processes can occur both within the bulk of the material and also at its surface. In devices where no special precautions have been taken to eliminate surface processes these will often be the dominant source of excess low-frequency noise. Since these processes do not contribute to the normal action of the device but rather cause, if anything, a deterioration in its signal performance they are clearly one of the parameters to be closely controlled, and if possible eliminated, in the design of low-noise devices. In bipolar transistors and junction field-effect transistors, where the surface plays no role in the amplification process, this can be achieved by careful surface treatment, but in insulated-gate field-effect devices, where the surface of the channel opposite the gate is critically involved in amplification, this has not so far been achieved. Although the flicker-noise performance of insulated-gate devices can be correlated with features of the channel surface (e.g. Sah and Hielscher, 1966), the devices are so much noisier than junction devices that they are not, at present, of much practical use in low-noise, low-frequency amplifiers.

Where carrier generation and recombination processes lead to currents in the external leads attached to a device they will result in excess noise components in these currents. Since the mean external current is generally proportional to the strength of the generation or recombination processes and this, in turn, determines the over-all magnitude of the excess noise power spectrum, the excess noise will be proportional to the mean current. Generation and recombination processes, however, can also cause excess noise, even if the carriers involved do not lead directly to external currents. This is, for example, the main excess noise mechanism in a junction field-effect transistor in which all surface noise effects have been eliminated. Within the channel and within the gate, where majority carrier conduction processes are dominant, the generation and recombination of minority carriers has little effect. However in the gate–channel depletion layer the appearance or disappearance of minority carriers has an appreciable effect on the potential distribution and produces a modulation of the channel current. This is equivalent in its effect to a fluctuating voltage applied to the gate. Thus in low-noise junction field-effect devices the main additional noise at low frequencies is equivalent to a noise-voltage generator in series with the gate connection.

In bipolar transistors minority carriers generated in the base produce insignificant effects, but the recombination of minority carriers injected into the base from the emitter is the cause of the direct base current. Fluctuations associated with this recombination rate lead to excess noise in the base current, and since the mean rate is proportional to the base current the noise power spectrum is also proportional to this current. The collector current is also proportional to the base current and is a somewhat more accessible variable, so that it is usual to regard the excess base current noise as proportional to the collector current.

In field-effect transistors the additional $1/f$ noise is equivalent to additional voltage fluctuations at the input (gate) terminal. Since for a given available input power the mean square signal input voltage is proportional to the source impedance, field-effect transistors give the best low-frequency signal to noise ratio with a high source impedance; generally at least 100kΩ is required. By way of contrast, the excess noise in a bipolar transistor appears as a fluctuating current in the input (base) lead, and so a low-impedance source, giving a large signal current for a given input power, is required. In fact, the optimum value is equal to the internal series base resistance of the device and is usually near 200 Ω. In this case the excess noise increases with increasing bias current but, as the white noise has a different current dependence, there is usually both an optimum source impedance and an optimum bias current. With suitable source impedances either type of device can be used to obtain low-noise, low-frequency amplification. We shall discuss this in more detail in later chapters. We may remark, however, that there is a very considerable difference between the low-frequency noise performance of 'run of the mill' devices and devices designed to minimize the excess noise.

In diodes all the current injected as minority carriers across a forward-biased junction recombines in the bulk material or at its surface. Excess noise associated with these processes is not usually of practical importance since, when diodes are used as detectors, the rectified signal always results in a mean bias in the reverse direction. The reverse current is critically dependent on minority carrier generation near the junction and fluctuations in the generation rate result in excess noise in the reverse current. At normal radio-frequencies diode detectors are usually only employed after appreciable radio-frequency gain has been achieved, and so noise generated in the diode is usually insignificant compared with the amplified noise from the input circuits. It is also unusual to find a diode mixer input stage used at these frequencies. Most microwave receivers however employ a diode mixer as the first stage, and noise generated in this diode as a result of the reverse bias applied by the local oscillator is significant. Since these diodes have a very small active cross-section and a large ratio of surface to volume and since control of surface conditions is difficult, the excess noise is large and may exceed the white noise output at frequencies as high as 100 MHz. For this if for no other reason, it is usually necessary to use a relatively high intermediate frequency in a microwave superhet receiver.

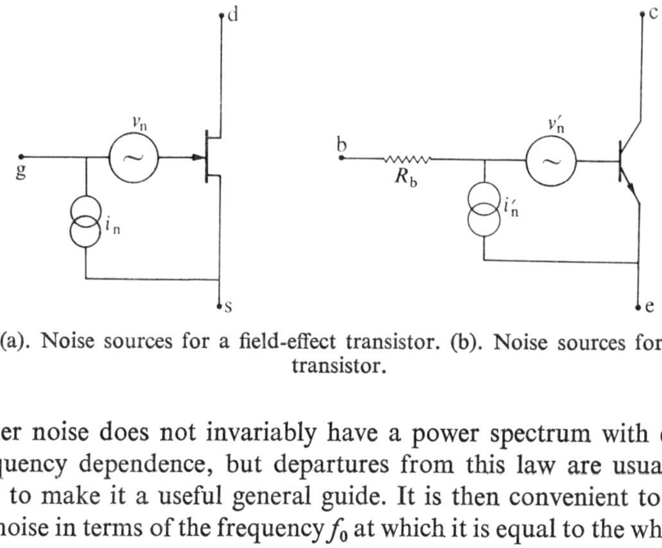

FIG. 7.1(a). Noise sources for a field-effect transistor. (b). Noise sources for a bipolar transistor.

Flicker noise does not invariably have a power spectrum with exactly a $1/f$ frequency dependence, but departures from this law are usually small enough to make it a useful general guide. It is then convenient to describe flicker noise in terms of the frequency f_0 at which it is equal to the white noise. Thus if we express the total noise power spectrum W in terms of the white-noise power spectrum w as

$$W = w(1+f_0/f), \tag{7.6}$$

the corner frequency f_0 gives a direct indication of the frequency range in which flicker noise becomes significant. The noise generators associated with a field-effect transistor and a bipolar transistor at moderate frequencies are shown in Figs. 7.1(a) and (b). The voltage and current generators have white power spectra w_v, w_i or w'_v and w'_i at moderately high frequencies. At low frequencies the current generator associated with the field-effect device is not changed, but w_v is increased so that

$$W_i = w_i, \qquad W_v = w_v(1+f_0/f). \qquad \text{(7.7 f.e.t.)}$$

For the bipolar transistor, on the other hand, we have

$$W_i' = w_i'(1+f_0'/f), \qquad W_v' = w_v'. \tag{7.7b.p.}$$

Note that in this case the current generator occurs to the right of the series base resistance R_b and so generates an input noise voltage $i_n R_b$ even if the external base lead b is shorted to the emitter lead e.

Although flicker noise is only directly manifest in low-frequency amplifiers, flicker noise in a non-linear radio-frequency circuit can produce noise in sidebands near the radio-frequency carrier. This is expecially troublesome in oscillators, which must be essentially non-linear circuits to produce oscillations of stable amplitude. The degree to which flicker noise modulates the output producing random amplitude, frequency, or phase modulation depends not only on the flicker-noise output of the active device but also on the circuit configuration. In general, flicker-noise effects are most significant in relatively low-level oscillators stabilized by a small curvature of the transfer characteristic of the active device; they are least significant in high-level oscillators stabilized by the presence of a sharp knee in the transfer characteristic.

Unlike shot noise and thermal noise, flicker noise does not lend itself to precise or detailed calculations. Those features of a device which determine the flicker-noise intensity, though sometimes accessible to sophisticated measurement, are not simply related to either gross features of the device design or to externally accessible variables or measurable device parameters. Thus the circuit designer is more than usually dependent on the information furnished by the device manufacturer. In some cases, e.g. the Mullard specifications for the field-effect device BFW11 or the p–n–p transistor BCY71, this is entirely adequate. Unfortunately, not every other manufacturer is so specific.

Although flicker noise generated in active devices is ultimately of most practical importance, any component subjected to either a d.c. bias or a strong a.c. signal can generate flicker noise. We have already mentioned composition resistors and ceramic capacitors (such as the low-voltage disc bypass types) which have an appreciable leakage current. This by no means exhausts the list. If a circuit displays an apparently inexplicable excess noise at low frequency, almost any component carrying either a direct or an alternating current is suspect. Contact resistance is a fruitful source of excess noise. This may occur in soldered, wrapped, or welded joints, and some of these joints may be within components. In one instance, in the author's experience, excess noise in a high-level nuclear quadrupole resonance spectrometer was eventually traced to, perhaps the most unlikely of all components, a 5 pF silver-mica capacitor in the radio-frequency tank circuit. The low-frequency modulation of the radio-frequency output of the oscillator due to contact noise in this capacitor was nearly double the noise due to all other sources.

It is also common to find that a substantial fraction of the excess noise in low-frequency amplifiers is due to the use of the wrong type of coupling capacitor.

Because of the existence of flicker noise the design of low-noise, very low-frequency amplifiers presents problems in many ways as intractable as those associated with very high frequencies. There is no doubt that, with the devices currently available, the design of low-noise amplifiers is easiest if the frequency of operation is above 10 kHz and less than 50 MHz.

At very low frequencies fluctuations due to flicker noise have to be considered in conjunction with slow drifts in the operating point of an amplifier, and steps taken to minimize one effect may aggravate the other. Thus, for example, whereas the significance of flicker noise in a field-effect transistor amplifier may be minimized by using a high-impedance source unfortunately this will increase the drift due to changing leakage currents in the input circuit. Equally, attempts to minimize the drift either by the use of additional components in a d.c. feedback loop or by the use of temperature-compensating elements may introduce additional flicker noise.

Any piece of electronic equipment requires a power supply to provide the necessary bias voltage, and the output of most power supplies contains appreciable noise components, expecially at low frequencies. Thus, for example, if a stabilized supply yields a nominal output V and maintains this output by using a differential amplifier to compare a fraction of the output with a reference voltage V/n, the mean square output noise voltage cannot be less than $n^2v_n^2$, where v_n^2 is the sum of the mean square fluctuations in the reference voltage and at the input to the differential amplifier. Power-supply noise is introduced easily into the input stages of an amplifier, and equally easily is mistaken for noise generated within the amplifier. For this reason very considerable care has to be taken with the design of the bias circuits for the input stages.

8 Noise in Vacuum Tubes

8.1. Introduction

ALTHOUGH for most purposes vacuum-tube amplifiers have been replaced by transistors, and indeed usually have inferior noise performance, nevertheless they are still occasionally used, and it is necessary to have some idea about their limitations. In addition, most of what we know about noise in electronics was originally discovered using vacuum tubes and, as a result, the classical literature of the subject is rather inaccessible without some knowledge of noise in vacuum tubes. We may add that, although the concepts developed in the course of analysing vacuum-tube noise do not have any very direct application to current semiconductor devices, there is no reason why they should not apply to devices yet to be discovered.

8.2. Noise in the space-charge limited diode

The power spectrum of the current in a temperature-limited diode carrying a mean current I_0 is given by the simple shot-noise formula $2eI_0$, and this can be verified readily by experiment. Experimentally, however, the noise in a space-charge limited diode, for the same current, is much less, in some cases by a factor of 1000. This remained a mystery for many years, and the literature of the 1930s is full of attempts at an explanation, most of them wholely erroneous and some unfortunately still current. One thing, however, is clear; this is that the smoothing occurs because of the effect of space charge. Indeed, on a naive picture we should expect, since the current is determined solely by the geometry of the diode, the applied anode voltage, and the values of the constants e, m, and ϵ_0, that the anode current should display no fluctuations, whatever the fluctuations in the emission from the cathode. This result would indeed be correct if all electrons were emitted from the cathode with the same velocity. It was the recognition that the thermal distribution of emission velocities played a significant role that gradually led to the formulation of a correct theory by Schottky (1937), Spenke (1937), and especially Thompson, North, and Harris (1940). The account that we shall give is based on the last named authors' work. The analysis is limited to a one-dimensional treatment of a plane parallel structure and neglects the variation of the potential and current over the cross-section of the system. It is confined also to low frequencies, i.e. below about 400 MHz, where the effects of transit time can be neglected. We shall deal with noise in electron beams at higher frequencies in Chapter 15 in connection with travelling-wave tubes.

Electrons from a cathode at a temperature T are emitted with a distribution of velocities and, for the moment, only the emission velocity normal to the cathode surface need concern us. If the energy associated with this component velocity is E and I_0 is the total emitted current, the current emitted in the range E to $E+\mathrm{d}E$ is

$$\mathrm{d}I = I_0 \exp(-E/kT)\,\mathrm{d}E/kT. \tag{8.1}$$

As we saw in Chapter 5, each component of the total current displays full shot noise and fluctuations in different components with different values of E are uncorrelated. The power spectrum associated with $\mathrm{d}I$ is therefore $2e\mathrm{d}I$.

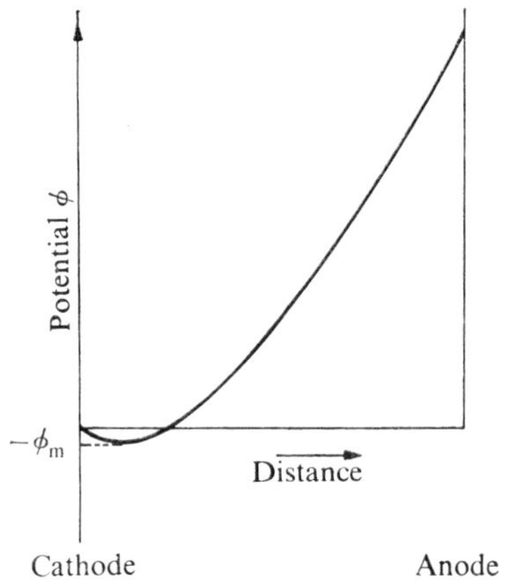

FIG. 8.1. Potential distribution in a space-charge limited diode.

In the region between the cathode and anode, the potential satisfies Poisson's equation $\partial^2\phi/\partial x^2 = -\rho/\epsilon_0$ and, since ρ is negative, a plot of ϕ against distance x from the cathode is concave upwards. Since, if $\partial\phi/\partial x$ is everywhere positive or zero, all electrons emitted will be accelerated to the anode and therefore the diode will not be space-charge limited, the potential must have a minimum as shown in Fig. 8.1. Electrons will only reach the anode if their initial energy E exceeds $e\phi_m$, and so the anode current is

$$I = \int_{e\phi_m}^{\infty} I_0 \exp\left(-\frac{E}{kT}\right)\frac{\mathrm{d}E}{kT} = I_0 \exp\left(-\frac{e\phi_m}{kT}\right). \tag{8.2}$$

Since in the region between the potential minimum and the anode the current is still controlled by space charge, i.e. the charge density due to the current must make $\partial\phi/\partial x$ zero at the minimum, there will be a unique value of ϕ_m for a given anode voltage, spacing, cathode temperature T, and total emission I_0; clearly also $e\phi_m = kT\log(I_0/I)$ and, in a typical case with $I_0 \sim 10I$ and $T \sim 1000$ K, this gives a minimum about 0·25 V deep, located a fraction of a millimetre from the cathode. If I_0 fluctuates the value of ϕ_m will also fluctuate, and a detailed calculation shows that an over-all fluctuation δI_0 in the total emission is almost entirely eliminated by the resulting change in ϕ_m. Flicker-noise fluctuations, which result from a general change in I_0 associated with a particular region of the cathode surface, in fact are smoothed very effectively. However, we have to deal with independent fluctuations in electrons emitted in different velocity, or energy, classes. A positive fluctuation in $\mathrm{d}I$ for the class of energy E has two effects. It adds directly to the anode current, and it depresses the potential minimum, thus turning back some electrons which otherwise might have reached the anode. These compensating fluctuations, it should be noted, will be drawn solely from electrons emitted with energy $E \sim e\phi_m$. If the additional electrons come from an energy class with $E < e\phi_m$ they will not reach the anode in any class, and so their sole effect, due to the depression of ϕ_m, will be a reduction in the anode current. Clearly, we must express the effect on the anode current of a fluctuation $\delta\{\mathrm{d}I(E)\}$ of energy E as $\gamma(E)\delta\{\mathrm{d}I(E)\}$, where $\gamma(E)$ depends on E and may be positive or negative. Since fluctuations in each energy class are uncorrelated, the fluctuation in the anode current is the sum of means of terms of the form $\gamma^2(E)[\delta\{\mathrm{d}I(E)\}]^2$ and, since the power spectrum of $\mathrm{d}I(E)$ is $2e\mathrm{d}I$, we have for the anode-current power spectrum

$$w(f) = \int 2\gamma^2(E)e\,\mathrm{d}I = \int_0^\infty 2\gamma^2(E)I_0\exp\left(-\frac{E}{kT}\right)\frac{\mathrm{d}E}{kT}.$$

The evaluation of $\gamma(E)$ and of this integral, in the general case, is exceedingly difficult but, if I is less than about $I_0/5$ and the anode voltage V_a satisfies $V_a > 30kT/e \sim 3$ V, the final result can be shown to be

$$w(f) = \Gamma^2 2eI, \tag{8.3}$$

where the smoothing factor is

$$\Gamma^2 = 9\left(1-\frac{\pi}{4}\right)\frac{kT}{eV_a}. \tag{8.4}$$

Thus, for example, if $V_a = 100$ V and $T = 1000$ K we have $\Gamma^2 \sim 2\times10^{-3}$, and the reduction in noise is most marked.

The fluctuations in the electron current leaving the potential minimum and reaching the anode consist of two components. First, there are the primary fluctuations which arise from cathode emission fluctuations in classes with $E > e\phi_m$, i.e. classes which reach the anode, and secondly there are the compensating fluctuations in classes with $E \sim e\phi_m$. These compensating fluctuations are correlated strongly with the primary fluctuations, which they almost cancel, but they also contain a weak component arising from fluctuations in the emission in classes with $E < e\phi_m$ which do not reach the anode.

Eqns (8.3) and (8.4) can be expressed in another, and more useful, form as follows. In a diode, in which all electrons are emitted with zero velocity, the anode current and voltage are related by Child's law $I = KV_a^{\frac{3}{2}}$ and, provided that eV_a is appreciably larger than kT, this result is not much affected by the thermal emission velocities. Thus the diode conductance can be expressed as

$$g = \frac{\partial I}{\partial V_a} = \frac{3}{2}\frac{I}{V_a}. \tag{8.5}$$

If this is used in (8.3) and (8.4) to eliminate I and V_a, we obtain

$$w = 3(1-\pi/4)4kTg \sim 0{\cdot}64(4kTg). \tag{8.6}$$

This is often expressed by saying that the current fluctuations are equal to Johnson noise in a conductance g at about $\frac{2}{3}$ of the cathode temperature. From time to time various abortive, and intrinsically fallacious, efforts have been made to find a thermodynamic derivation of this result. It is quite clear that a diode with a hot cathode, cold anode, and a power supply connected is not a system in thermodynamic equilibrium. Despite this, eqn (8.6) is useful. The equivalent circuit of a diode is a conductance g in parallel with a current generator of mean square strength $I_n^2 = 0{\cdot}64\ (4kTg\,\mathrm{d}f)$, where $\mathrm{d}f$ is the relevant bandwidth. In circuit calculations it can be treated as a passive conductance g at a temperature $0{\cdot}64\ T$, i.e. about twice room temperature. We can also use the result to express Γ^2 in terms of g and I. If $T \sim 1000$ K, typical of an oxide cathode, we have $\Gamma^2 \sim g/9I$, where g is in A V^{-1} (reciprocal ohms) and I in amperes.

These results, we should add, are not very well verified using readily available diodes. The residual gas leads to positive ion formation and, since these ions are formed randomly and tend to be trapped at the negative potential minimum, they engender excess noise. To verify eqn (8.4), for example, one requires a diode pumped to a much better vacuum than is necessary for other applications. In a triode or multigrid tube, however, the presence of the negative control grid diverts the ions away from the potential minimum and, in these tubes, as we discuss in the next sections, the consequences of eqn (8.4) are verified with reasonable accuracy.

8.3. Triodes

The a.c. behaviour of a triode can be analysed in terms of an equivalent diode whose anode plane coincides with the plane of the triode grid and whose effective voltage is $V_d = V_g + (1/\mu)V_a$, where V_g is the grid voltage, V_a the anode voltage, and μ is the amplification factor, which is determined by the tube geometry. The anode current is given by $I_a = KV_d^{\frac{3}{2}}$, and so the mutual conductance is

$$g_m = \frac{\partial I_a}{\partial V_g} = \frac{3}{2}\frac{I_a}{V_d}.$$

The noise power spectrum of the anode current is $2e\Gamma^2 I_a$, and we have $\Gamma^2 \sim g_m/9I_a$, so that

$$w = \frac{g_m}{9I_a}.2eI_a \approx \frac{eg_m}{4}. \tag{8.7}$$

A noise voltage with a power spectrum w/g_m^2 applied at the grid would produce identical output fluctuations, and it is convenient to refer the noise to this input terminal, where it can be compared directly with noise associated with the signal input. Since a resistance R_n at a temperature T_0 has a voltage spectrum $4kT_0R_n$ the noise due to the tube is equal to noise in a resistance given by

$$R_n = \frac{e}{4g_m}.\frac{1}{4kT_0}. \tag{8.8}$$

If T_0 is taken as 293 K, the usual reference temperature, we have $kT_0/e \sim 25$ mV, and so

$$R_n = \frac{2.5}{g_m}. \tag{8.9}$$

An equivalent circuit which represents noise in a triode amplifier connected to a signal source is shown in Fig. 8.2. The resistance R_n

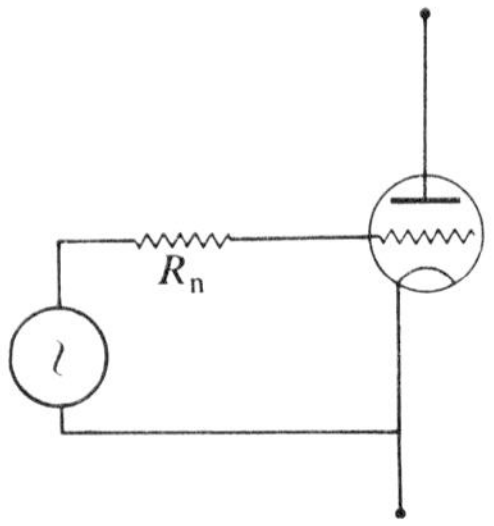

FIG. 8.2. Noise equivalent circuit for a triode.

serves solely to represent a noise voltage generator of power spectrum $4kT_0R_n$ and plays no other role in circuit equations. To see why this is a useful representation consider Fig. 8.3, in which a signal source of internal impedance R_s at T_s is connected to the triode. In the absence of other sources of noise the total mean square noise voltage acting on the (otherwise noiseless) triode is the sum of the Johnson noise in R_s and R_n, and its power spectrum is $4kT_sR_s+4kT_0R_n$. The mean square input noise is greater than the noise due to the source alone by a factor $F = 1+R_nT_0/R_sT_s$, and if, as is often the case, $T_s = T_0$ we have the very simple expression $F = 1+R_n/R_s$ for the increase in noise. The factor F is, as we shall see later, the noise figure for this particular amplifier.

In practice, eqn (8.9) is found to be accurate as a representation of noise to within about 20 per cent for most triodes over the frequency range from 10 kHz to 10 MHz. At lower frequencies flicker noise must be considered, and at higher frequencies the noise increase due to transit-time effects (see § 8.5) must be taken in account. We note that, since triodes with g_m of the order of 25 mA V^{-1} are available, the value of R_n need only be 100 Ω. If the source impedance is 10^5 Ω, the degradation of the noise performance due

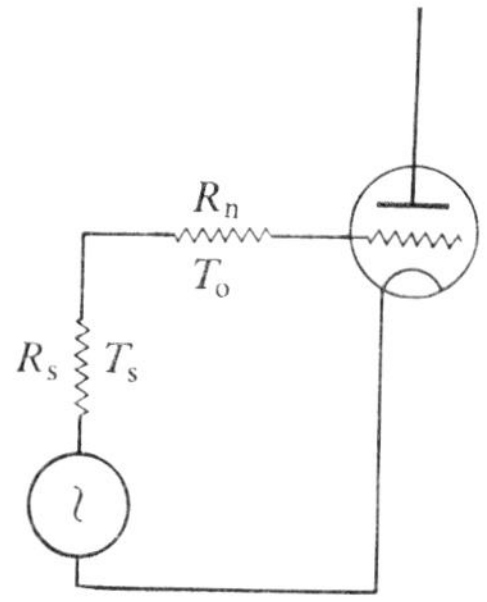

FIG. 8.3. A triode connected to a signal source of internal impedance R_s at T_s.

to shot noise is only 1·001, which is negligibly different from unity. In such an amplifier virtually all the output noise derives from the amplified Johnson noise of the source impedance.

8.4. Multigrid tubes–partition noise

Even in tube circuits, triodes are relatively rare, their amplification factor is low, usually less than 100, and the Miller effect, due to anode to grid capacitance, tends to make triode amplifiers unstable at relatively low radio-frequencies, i.e. around 1 MHz. Much the commonest vacuum tubes are the tetrode and its minor modification, the pentode. In these tubes a positive second grid between the negative signal grid and anode controls the d.c. bias condition, and the anode serves solely to collect electrons passing through the second grid. The anode current serves as the output signal and, since changes in anode voltage have little influence on the current, the amplification factor is large. The second grid, between the control grid and the anode, reduces the Miller effect to negligible proportions. Because, however, the current that reaches the anode is the result of a random division of the cathode current at the second grid, there is an additional source of noise, known as partition noise.

The current approaching the second grid is smooth because the primary shot-noise fluctuations are compensated by correlated fluctuations. The primary and compensating fluctuations, however, consist of different groups of electrons, and there is no correlation between their transverse velocities or trajectories. Thus each group of fluctuations divides randomly and independently at the second grid. A primary fluctuation may hit the grid, and its compensating fluctuation miss it, or vice versa. A quantitative calculation (Thompson, North, and Harris 1940) shows that the effective space-charge smoothing factors for the anode and second grid currents are

$$\Gamma_a^2 = 1-\alpha(1-\Gamma^2) \tag{8.10a}$$

and

$$\Gamma_s^2 = \alpha+(1-\alpha)\Gamma^2, \tag{8.10b}$$

where Γ^2 is the initial factor and α the fraction of the cathode current reaching the anode. The additional fluctuations in the two currents are completely correlated and, if the two electrodes are connected and the tube used as a triode, its noise performance is that of a triode. Eqn (8.10a) can be written as

$$\Gamma_a^2 = \frac{I_2}{I_c}+\frac{I_a}{I_c}\Gamma^2 \tag{8.11}$$

in terms of the screen-grid, anode, and cathode currents. It follows that the equivalent noise resistance of a pentode or tetrode is

$$R_n = \frac{2.5}{g_m}\left(1+9\frac{I_2 I_c}{I_a g_m}\right), \tag{8.12a}$$

where g_m is the total transconductance $\partial I_c/\partial v_g$. In terms of the anode transconductance $g_a = \partial I_a/\partial v_g = g_m I_a/I_c$, we have

$$R_n \approx \frac{2.5}{g_a}\frac{I_a}{I_c}\left(1+9\frac{I_a}{g_a}\right). \tag{8.12b}$$

If, for example, $I_2 = 1$ mA, $I_a = 9$ mA, and $I_c = 10$ mA, while $g_m = 5$ mA V^{-1}, the factor in the bracket is 3 and, in practice, pentodes and tetrodes are generally about 3 times as noisy as a triode of the same mutual conductance. Although the reasoning which leads to (8.12) is crude it represents noise in multigrid tubes to within about 20 per cent accuracy.

We note that eqn (8.10b) indicates that as α approaches unity the space-charge smoothing factor for the second grid current also approaches unity whatever the value of Γ^2. This is a general result. If an electrode intercepts only a small fraction $1-\alpha$ of a larger current, the current reaching the electrode displays full shot noise. For this reason one should be careful to avoid using positive grids as signal-output electrodes in low-noise systems (unless, of course, all the positive electrodes are strapped together).

8.5. Higher frequencies and induced grid noise

As electrons leave the cathode and pass towards, and through, the grid on their way to the anode, they induce a pulse of current, initially negative and then positive, in the grid circuit. The net charge in each pulse is zero but, if the period of the signal voltage applied to the grid is comparable with the transit time, the effect extracts energy from the input circuit and leads to the appearance of a finite input, or grid to cathode, conductance in parallel with the input capacitance. The magnitude of this conductance at a frequency f is of the order of

$$G_\tau \sim g_m(2\pi f\tau)^2/20 \sim 2g_m f^2\tau^2, \tag{8.13}$$

where τ is the cathode to grid transit time. Thus, if $\tau = 10^{-9}$ s and $g_m = 20$ mA V^{-1} the input conductance at 100 MHz is 0·4 mA V^{-1}. The electrons, in

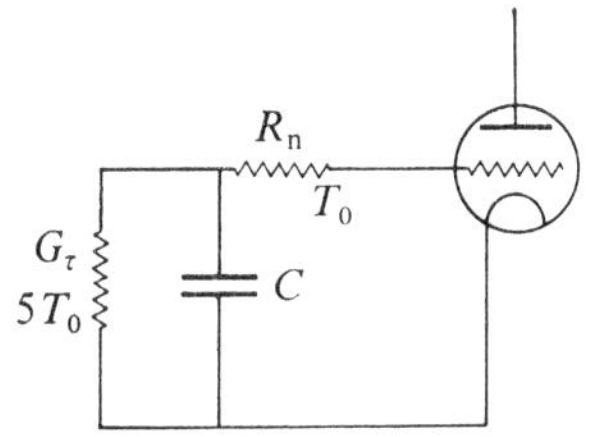

FIG. 8.4. High-frequency noise sources for a triode.

their passage, also induce noise current in the grid circuit and, although a detailed calculation is difficult, the power spectrum of this current is also proportional to g_m and $(2\pi f\tau)^2$. It is approximately (North and Ferris, 1941) given by

$$w = \frac{20}{3}\left(1-\frac{\pi}{4}\right)4kT_cG_\tau. \tag{8.14}$$

If T_c, the cathode temperature, is 1000 K, this is equivalent to Johnson noise in a conductance G_τ at 1450 K, which is about $5T_0$ where T_0 is the standard reference temperature 293 K. The high-frequency equivalent input circuit of a triode, including the transit-time conductance G_τ and the input capacitance C as well as the normal noise resistance R_n, is shown in Fig. 8.4. Note that R_n is to the right of C and G_τ.

In terms of the input admittance $Y = G_\tau + j\omega C$, and voltage and current generators, the circuit is as shown in Fig. 8.5, and the power spectra of the two noise sources are

$$w_v = 4kT_0R_n = 10kT_0/g_m, \tag{8.15a}$$

$$w_i = 20kT_0G_\tau = 20kT_0/R_\tau. \tag{8.15b}$$

Manufacturers usually give the value of R_τ, the inverse of G_τ, at a specified frequency. Thus, for the pentode type E180F, we have

$$R_\tau = 1{\cdot}5\times10^{19}\,\mathrm{f}^{-2}\,\Omega.$$

At 100 MHz this is only 1500 Ω. In general, $R_r g_m$ is much the same for all small-signal pentodes, indicating that the cathode to grid transit times are similar.

The noise voltage and current generators in Fig. 8.5 are partially correlated, since they arise from the same fundamental shot-noise process. The degree of correlation is small, and the correlated components are in quadrature. This, as we shall see in Chapter 12, means that the effects of the correlation are very small. Although, in the past, attempts to use this correlation to improve the noise performance of high-frequency amplifiers have provided ample

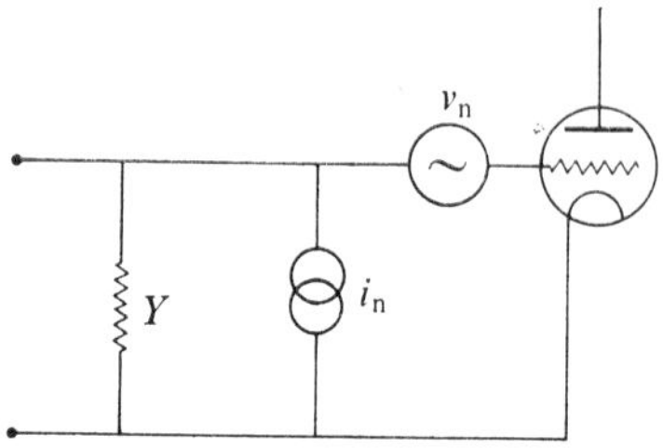

FIG. 8.5. Input parameters for a triode.

opportunities for virtuoso displays of elementary algebra, they have led to very little significant improvement. Unfortunately, similarly insignificant correlations occur in both bipolar and field-effect transistors, and these have been exploited similarly, with little practical advantage.

8.6. Ion or grid-current noise

The vacuum in electron tubes is not perfect, and residual gas is ionized by the electrons producing positive ions which drift to the grid and give a finite grid current I_g. This current, as we might expect, displays the full shot-noise power spectrum

$$w = 2eI_g. \tag{8.16}$$

However, it will be important only if the resistance R in the grid circuit is high, so that the voltage fluctuations $2eI_gR^2\,\mathrm{d}f$ due to ion noise exceed the Johnson noise $4kTR\,\mathrm{d}f$ due to R. This occurs when

$$eI_gR > 2kT_0$$

or

$$I_gR > \tfrac{1}{20}. \tag{8.17}$$

Thus, if I_g is of the order of 10^{-9}–10^{-7} A, which is the case in most small-signal tubes, ion noise is only significant if R is of the order of megohms. Special tubes with a better vacuum and I_g as low as 10^{-11} A are manufactured for use in electrometers, and then values of R in excess of 10^9 Ω are tolerable. If an electrometer-triode is used to measure a current I flowing in a resistance

R, the signal to noise ratio with a bandwidth f is given by

$$\left(\frac{S}{N}\right)^2 = \frac{R^2 I^2}{4kTR\,\Delta f + 2eI_g R^2\,\Delta f},$$

and, as R is made very large, this tends to the limiting value $I^2/2eI_g\Delta f$. If, for example, $I_g = 10^{-11}$ A and $f = 1$ Hz, a current of $1{\cdot}8 \times 10^{-15}$ A gives unity signal to noise ratio. Since the required value of R is about $10^{10}\ \Omega$ the time constant of the circuit for an input capacitance of 10 pF is 0·1 s.

When the grid potential is not very negative with respect to the cathode, the most energetic electrons emitted by the cathode can reach the grid. Since fluctuations in the emission of these very energetic electrons are not smoothed by space charge, the resulting grid current displays full shot noise. It can be a significant source of noise in some applications.

8.7. Summary

As we indicated at the beginning of this chapter, the study of noise in vacuum tubes is mainly of historical interest as a background to the literature on noise prior to about 1965. Triodes and pentodes have been superseded by transistors as low-noise amplifiers at all frequencies up to about 5000 MHz. which is the upper limit for both types of device, and electrometer-triodes have been replaced by field-effect transistors. Triodes and pentodes are now only important in high-power radio-frequency amplifiers and oscillators. Nevertheless, ideas drawn from the study of vacuum-tube noise continue to influence discussions of transistor noise and, it may be added, are often applied erroneously. Thus, for example, because of a superficial resemblance to triode noise, noise in field-effect transistors was thought to be a smoothed shot-noise process. The correct interpretation (van der Ziel, 1963) is as a Johnson-noise process. It is therefore necessary, even in dealing with transistors, to have some understanding of vacuum-tube noise. Fortunately, as we shall see, noise in semiconductor devices is altogether simpler to analyse.

9 Noise in Junction Diodes

A major clarification of our understanding of noise in semiconductor devices resulted from the publication of three important papers by van der Ziel and Becking (1958) and van der Ziel (1962, 1963), and our treatment of these topics in this and the next two chapters is based on their ideas, although the reader should consult the note at the end of this chapter.

9.1. Introduction

A junction diode consists of two contacting regions of relatively highly doped and conducting material, of two different types, both regions being fitted with ohmic contacts. If the conductivity of the two bulk regions is high enough, a potential difference applied to the ohmic contacts produces almost no electric field in these regions, and the entire potential difference appears across the depletion layer at the junction. We now show that this implies that the noise observed in the external leads is entirely associated with the motion of charge carriers in the depletion layer. Except for a small Johnson- or thermal-noise contribution due to their finite conductance, the bulk p and n regions make no contribution to the noise.

If two conducting electrodes, in our case the leads and contacts, are maintained at a potential difference V by a battery, producing an electric field $\mathbf{E}$ in their vicinity, and a particle of charge q and velocity v moves in this region, the field does work on the particle at a rate $q\mathbf{E}.\mathbf{v}$. This energy must be supplied by the battery, and so a current $I = (q/V)\mathbf{E}.\mathbf{v}$ flows in the external circuit. Since $\mathbf{E}V^{-1}$ is purely geometric factor this is an expression for the current induced in the external lead by the motion of the particle (see Appendix A for a fuller discussion). It vanishes for motion in a region where $\mathbf{E} = 0$. Further, if in a time t the particle moves from a position where the potential is ϕ_1 to a position where it is ϕ_2, the charge δq which flows in the external circuit is

$$\delta q = \int I\,\mathrm{d}t = \frac{q}{V}\int \mathbf{E}.\mathbf{v}\,\mathrm{d}t = \frac{q}{V}\int_{\phi_1}^{\phi_2} -\nabla\phi.\mathrm{d}\mathbf{r} = -q\frac{\phi_2-\phi_1}{V}.$$

The sign in this relation need not concern us, it is purely conventional and we can replace this result by the statement: when a charged particle q traverses a region in which the potential change is a fraction α of the applied voltage, the charge transferred in the external circuit is αq. For a complete traversal from one contact to the other we have $\alpha = 1$ and a rather obvious result.

Applied to a junction diode this result means that, when a carrier moves from one region to the other, the current in the leads consists of a short pulse, of total charge q, which occurs as the carrier traverses the depletion layer. The motion of the particle prior to, and after, this transit, i.e. its motion in the n and p regions, produces no external effect other than that described by the Johnson noise associated with these regions.

The motion of carriers in a depletion layer, where they make no significant thermalizing collisions, fulfils all the criteria established in Chapter 5 for a shot-noise process. Thus if, in a p–n junction diode under forward bias, all the current is carried by electrons injected into the p region from the conduction band of the n region, the power spectrum associated with a mean current I is simply $2eI$. We now have to see how this result is modified by other processes and how it varies with bias and frequency.

In elementary discussions of the p–n junction it is assumed that the current consists of two distinct and independent components: electrons being exchanged between the two conduction bands and electrons being exchanged between the two valence bands. Usually the latter process is described as hole transfer. We shall begin by considering this model, but later we shall go on to consider a third process which, as it happens, is significant in silicon diodes with reverse, zero, or low forward bias, and plays an unwanted role in bipolar transistors. In this process electrons change bands within the depletion layer. The mechanism involves an intermediate recombination centre, or trap, within the depletion layer, and since its net effect is either the generation of a hole–electron pair or the recombination of a pair the current that results is known as recombination current.

9.2. The ideal diode

In the elementary case, in which recombination is neglected, electron and hole processes are independent and additive, both in their effects on the mean current and on the noise. We need consider therefore only one process, and for clarity we choose electrons. The electron current consists of two components: a forward component consisting of electrons from the n region crossing to the p region and a reverse component of electrons passing from p to n. At zero bias these currents balance on average. The reverse current is controlled only by the density of electrons (minority carriers) in the p material, and there is no barrier to their transit to the n region. This current is unchanged by an applied bias. We denote it by I_{d0}. The forward current is controlled partially by the majority carrier (electron) concentation in the n material but, since this is large, the dominant effect is due to the retarding potential barrier at the junction. An applied potential V, with the p region positive, reduces this barrier, and the forward current increases by a factor $\exp(eV/kT)$. Thus the total current is

$$I = I_{d0}\{\exp(eV/kT) - 1\}. \tag{9.1}$$

Each component of I displays full shot noise, and so the power spectrum of the current noise is

$$w = 2eI_{d0}\{\exp(eV/kT)+1\}. \tag{9.2}$$

At large forward bias, where the forward current dominates, this reduces to

$$w = 2eI. \tag{9.3}$$

For large reverse bias, where $|I| = I_{d0}$, the power spectrum is the same. The differential, or small-signal, conductance of the diode is

$$G = \frac{dI}{dV} = \frac{eI_{d0}}{kT}\exp\left(\frac{eV}{kT}\right), \tag{9.4}$$

and we can use this to write the general expression (9.2) as

$$w = 4kTG - 2eI. \tag{9.5}$$

At zero bias the diode is a system in thermodynamic equilibrium (contrast a vacuum diode), and it is therefore reassuring that when I (the net current) is zero we get the Nyquist result $w = 4kTG$.

We have neglected so far the effects of electron transit time in the depletion layer but, since the electron velocities are of the order of 10^5 m s^{-1} and the layer-width is of the order of 10^{-6} m, this time is only about 10^{-11} s. At any practical frequency, transit-time effects are negligible or can be made so by a proper choice of doping profile to reduce the width of the depletion layer. A finite transit time reduces the noise at high frequencies but also reduces the signal response and, in general, leads to an over-all deterioration in performance. There is, however, one frequency-dependent effect which we must consider. An electron constituting part of the forward current is injected into the p material, where it is a minority carrier, and the presence of an excess of injected minority carriers appreciably affects their concentration and the reverse current. This effect may be described by saying that there is an appreciable probability that an injected minority carrier will return across the junction to its starting point in the n material. This effect does not occur to any appreciable extent for electrons arriving in the n material, for the electron concentration there is already large. We have then to consider only the consequences of this process in the forward current. We begin by considering a component $I_{\tau 0}\exp(eV/kT)$ of the forward current which returns across the junction after a definite delay τ. This contributes a term

$$\Delta Y = \Delta G + j\,\Delta B = \frac{dI}{dV}(1 - \exp(-j\omega\tau))$$

$$= \frac{eI_{d0}}{kT}\exp\left(\frac{eV}{kT}\right)(1 - \cos\omega\tau + j\sin\omega\tau)$$

to the diode admittance, and the associated noise power spectrum is

$$\Delta w = |1-\exp(-j\omega\tau)|^2\, 2eI_{\tau 0}\exp\left(\frac{eV}{kT}\right) = 4eI_{\tau 0}\exp\left(\frac{eV}{kT}\right)(1-\cos\omega\tau).$$

This can be expressed as

$$\Delta w = 4kT\,\Delta G \tag{9.6}$$

and, since this relation is independent of τ, it also holds for a random distribution of dwell times. If this result is combined with eqns (9.3) and (9.5) we see that the noise can be expressed as

$$w(\omega) = 4kTG(\omega)-2eI \tag{9.7}$$

and that, for appreciable forward bias,

$$w(\omega) = 2eI+4kT(G(\omega)-G(0)). \tag{9.8a}$$

Since

$$w(0) = 2eI+4eI_{d0} = 2kTG(0),$$

we can also express $w(\omega)$ as

$$w(\omega) = 2kTG(0)+4kT(G(\omega)-G(0)). \tag{9.8b}$$

We see from (9.8b) that at moderate frequencies, where we can ignore the difference between $G(\omega)$ and $G(0)$, a forward-biased diode generates half the noise generated by a equal passive conductance at the same temperature. Thus if a two-terminal circuit, all at a temperature T, contains a forward-biased diode the available noise power at its terminals may be less than kT per unit frequency interval. The diode appears to act as a noise-sink, or heat-pump, absorbing thermal-noise power from the circuit and delivering it as heat to the surroundings. This violates no law of thermodynamics, since the biased diode is not a system in thermal equilibrium and work is being done continuously on the system by the source of the bias. The system is analogous to a domestic refrigerator. As long as this is biased by connection to the mains it can transfer heat from the ice compartment to the ambient air, even though this is at a higher temperature.

This result illustrates two general points. First, if a system, though at a uniform temperature, is not in a state of thermodynamic equilibrium, the available noise power at its terminals is not solely determined by its temperature; and secondly, we must beware of applying thermodynamic arguments to systems which are not in internal thermodynamic equilibrium, even if the non-equilibrium components of the system are not 'active' in the sense that they can amplify coherent signals.

9.3. Recombination effects

According to eqn (9.1) all diodes have the same form of I/V characteristic and if, for example, the reverse current I_{d0} is 10^{-9} A and $T = 300$ K the forward current when $V = 0{\cdot}6$ V should be over 10 A. Anyone familiar with the behaviour of real diodes will know that this is not the case; in particular, in silicon diodes a reverse current of 10^{-9} A is more likely to accompany a forward current of 10^{-3} A at 0·6 V and 10^{-6} A rather than 0·5 A at 0·5 V. Thus the ideal model of the diode is not a very good description of real diodes. The reason is that we have neglected carrier recombination and generation in the depletion layer. In silicon diodes this process dominates the reverse characteristic and the behaviour at low forward bias.

We suppose, to start with, that there is a localized group of traps or recombination centres in the depletion layer at a point where, as the applied voltage varies by V, the potential varies by αV. These traps both serve to facilitate electron–hole recombination and to generate electron–hole pairs, i.e. to transfer electrons from the conduction band to the valence band and vice versa. If an electron–hole pair is created in the depletion layer the field sweeps the electron towards the n region and the hole towards the p region. This process contributes to the reverse current and, since its rate is controlled by the characteristics of the trap levels—the thermal energy available and the *electron* concentration in the *valence* band and the *hole* concentration in the *conduction* band, both of which are large—it does not vary much with voltage. This only alters the electron concentration in the conduction band and the hole concentration in the valence band. By contrast, the recombination process, which transfers an electron from the conduction band to the valence band, depends on the electron concentration in the conduction band and the hole concentration in the valence band, and therefore is voltage-dependent. We therefore expect an expression for the current of the general form $I_0\{\exp(\alpha eV/kT)-1\}$. Unfortunately, the situation is rather more complicated than we have suggested (see Sah, Noyce, and Shockley (1957) for a full discussion) because, with a finite number of recombination centres available, the two processes interfere. The final expression for the net current due to recombination in this group of traps is of the form

$$I_{r\alpha} = K'\frac{pn-n_i^2}{n+p+a'}, \tag{9.9}$$

where K' and a' are constants, p and n are the hole and electron concentrations at the location of the recombination centres, and $n_i = p_i$ is the electron, or hole, concentration in intrinsic material. With no applied voltage $pn = n_i^2$ and the two terms, pn representing forward current and n_i^2 representing reverse current, cancel in the numerator. For forward bias, n grows as $\exp\{\alpha eV/kT\}$ and p like $\exp\{(1-\alpha)eV/kT\}$, since αV and $-(1-\alpha)V$ are the changes in potential of the centres relative to the n and p regions respectively. Thus eqn (9.9) can be rewritten as

$$I_{r\alpha} = K\frac{\exp\{eV/kT\}-1}{\exp\{\alpha eV/kT\}+\exp\{(1-\alpha)eV/kT\}+a}. \tag{9.10}$$

Each component of I_r has full shot noise, and so the power spectrum is

$$w_r = 2eK\frac{\exp\{eV/kT\}+1}{\exp\{\alpha eV/kT\}+\exp\{(1-\alpha)eV/kT\}+a}. \tag{9.11}$$

The differential conductance at zero bias is easily shown to be

$$G_r = \frac{eK}{kT(2+a)},$$

and also at zero bias,

$$w_r = \frac{4eK}{2+a},$$

so that we confirm that when the system is in thermal equilibrium

$$w_r = 4kTG_r.$$

The value of the reverse current when V is negative and large is $-K/a$, and so we can express (8.10) in terms of $I_{r\alpha 0}$ as

$$I_{r\alpha} = I_{r\alpha 0}\frac{\exp\{eV/kT\}-1}{\exp\{\alpha eV/kT\}+\exp\{(1-\alpha)eV/kT\}+a}. \tag{9.12}$$

The value of α lies between 0 and 1, and the value of a in (9.12) depends on the energy of the recombination centres and is likely to be large, say 10^3–10^6 but not much larger. Thus for small forward bias we have

$$I_{r\alpha} \sim \frac{I_{r\alpha 0}}{a}\exp\left(\frac{eV}{kT}\right),$$

but as V is increased until either $\exp\{\alpha eV/kT\}$ or $\exp\{(1-\alpha)eV/kT\}$ exceeds a, which even if $\alpha = \frac{1}{2}$ will only require about 0·3 V, the relation becomes either

$$I_{r\alpha} \sim I_{r\alpha 0}\exp\{\alpha eV/kT\}$$

or

$$I_{r\alpha} \sim I_{r\alpha 0}\exp\{(1-\alpha)eV/kT\},$$

depending on whether α is less or greater than $\frac{1}{2}$. Even for one localized group of recombination centres the d.c. characteristic is quite complicated and very far from the simple relation (9.1). Clearly, when we attempt to average (9.12), which (Sah, Noyce, and Shockley, 1957) is in any case an approximation, over all positions of the recombination centres in the depletion layer, the resulting expression well be even more complicated. Nevertheless its consequences are relatively simple. The diode current can be expressed as a sum of a diffusion term and a recombination term as

$$I = I_{d0}\left\{\exp\left(\frac{eV}{kT}\right)-1\right\}+I_r. \tag{9.13}$$

For reverse bias and low forward bias the recombination term I_r, with its complex dependence on voltage, dominates, but eventually at higher forward bias, the diffusion term, with the largest exponent eV/kT compared with a maximum of $eV/2kT$ in I_r, takes over and is the dominant term. Moreover, and this is of crucial significance in transistors, the forward diffusion term is the only term associated with minority carrier injection.

The situation with respect to noise is simple only for either zero bias or forward and reverse biases where either the reverse or the forward current is negligible. At zero bias $w = 4kTG$, and for appreciable bias of either sign $w = 2eI$. Fortunately, the intermediate range is of no great interest. It will be obvious also that, in the presence of recombination where there is a time delay in the recombination process, the high-frequency response and noise will be exceedingly complex. The only simple situation occurs at large forward bias, where the forward diffusion current $I_{d0}\exp(eV/kT)$ is the only significant term in the current.

We have spent some time discussing the effects of recombination, not so much because it is important as because, in discussing transistor noise, we (correctly) base our results on the simple diode formula (9.1), whereas it is a matter of trivial observation to see that this formula does not describe the behaviour of a real silicon diode. It is important to know that this discrepancy is due to recombination and to recognize that recombination does not lead to minority carrier injection. It therefore plays no part in the basic mechanism which leads to transistor action.

9.4. Diode noise in circuits

The equivalent small-signal circuit of a diode including noise is shown in Fig. 9.1, where $Y = G+jB$ is the differential admittance of the diode and the power spectrum of i_n depends on the bias conditions.

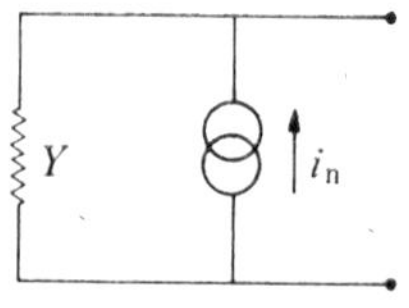

FIG. 9.1. Noise equivalent circuit for a diode.

The open-circuit noise voltage v_n is i_n/Y. If the diode is forward-biased, Y will be large and v_n small. Thus, for large forward bias, $Y = eI/kT$ and $w_i = 2eI$, so that the power spectrum of v_n is

$$w_v = 2\frac{(kT)^2}{eI} = \frac{2kT}{G}. \tag{9.14}$$

For example, if $I = 1$ mA, the effective resistance $1/G$ at room temperature is 25 Ω, and w_v is equal to Johnson noise in a 12·5 Ω resistor. The available power per unit bandwidth is $\frac{1}{2}kT$ and, provided I is large, independent of I. For reverse bias G is small, the available power is large and proportional to I, and in some cases the diode can be used as a variable noise source. It is also a highly undesirable circuit element in low-noise amplifiers. In general, diode noise is of little practical interest, since diodes are rarely used in the early stages of electronic systems. An exception is the microwave mixer diode and, even in this case, its main effect on the over-all noise performance is due mainly to its conversion loss (of signal) and its excess flicker noise. The fundamental shot-noise processes in the diode have a rather small effect.

Note added in proof

Since this chapter was written Buckingham and Faulkner (1974) have given a new analysis of noise in diodes and transistors. Although, for most practical purposes, their results are almost the same, the physical basis of their treatment is quite different and, to the author, more realistic. A brief account of their work is therefore included as Appendix B, and this should be read in conjunction with Chapters 9, 10, and 13.

10 Bipolar Transistors

10.1. The basic results

The behaviour of transistors, especially at high frequencies, is complicated by the presence of parasitic circuit elements, notably the series resistance R_b associated with the base and the capacitances associated with the two junctions. If, however, we neglect these elements the basic results are quite simple. Because n–p–n and p–n–p devices behave in essentially the same way, for clarity, we shall refer only to n–p–n devices.

Fig. 10.1 serves to define sign conventions for the d.c. currents I_e, I_b, I_c and the small-signal or noise currents i_e, i_b, and i_c. Currents in the three leads

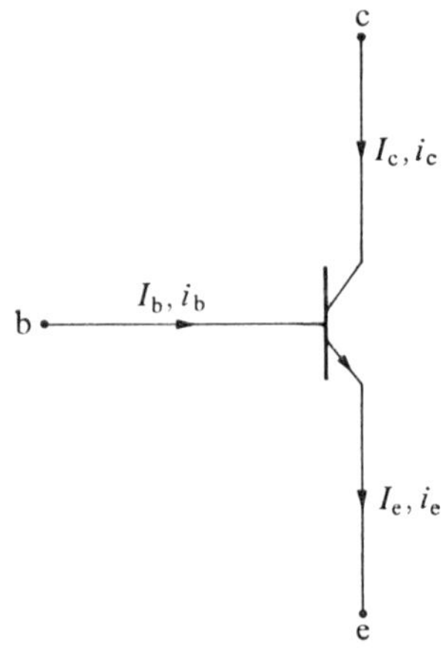

FIG. 10.1. Currents in the external leads of a bipolar transistor.

consist of pulses of charge, as electrons traverse the depletion layers, and so (see Chapter 5) they display shot noise. In particular, the collector current I_c, arising as it does from a succession of random processes, displays full shot noise, and its power spectrum is

$$w_c = 2eI_c. \tag{10.1a}$$

The emitter current I_e also displays full shot noise, but the emitter-current fluctuations also contain a component due to electrons making multiple transits across the depletion layer, thus its power spectrum is

$$w_e = 2eI_e + 4kT\,\Delta G_e, \tag{10.1b}$$

where G_e is the extra conductance of the emitter-base diode due to multiple transits. (But see Appendix B).

Each electron in the collector current was originally a component of the emitter current. A Fourier component i_c of the collector current at a frequency f would, if all the electrons took the same time τ to traverse the base, be completely correlated with a component of the emitter current at an earlier time, and we should have

$$\langle i_e i_c^* \rangle = \langle i_c i_c^* \rangle \exp(2\pi jf\tau), \text{ or } \langle i_e^* i_c \rangle = \langle i_c^* i_c \rangle \exp(-2\pi jf\tau).$$

Since the electron transit times are not all the same we have to average overall possible values of τ, and thus the correlation power spectrum w_{ec} associated with $i_e^* i_c$ is

$$w_{ec} = w_c\,\overline{\exp(-2\pi jf\tau)} = 2eI_c\,\overline{\exp(-2\pi jf\tau)}. \tag{10.2}$$

A signal component i_e at low frequencies leads to a signal component $i_c = \alpha_0 i_e$ and, at high frequencies, to a signal component

$$i_c = \alpha_0\,\overline{\exp(-2\pi jf\tau)}\,i_e;$$

the high-frequency current transfer factor is therefore

$$\alpha(f) = \alpha_0\,\overline{\exp(-2\pi jf\tau)}, \tag{10.3}$$

and so

$$w_{ec} = 2eI_c\alpha(f)/\alpha_0. \tag{10.1c}$$

Fluctuations in the base current are related directly to those in the emitter and collector currents, since $i_b = i_e - i_\alpha$, and so

$$w_b = w_e + w_c - w_{ec} - w_{ec}^* = 2eI_e + 2eI_c\left(1 - \frac{\alpha(f)+\alpha^*(f)}{\alpha_0}\right) + 4kT\,\Delta G_e,$$

or, since $I_b = I_e - I_c$,

$$w_b = 2eI_b + \left(2 - \frac{\alpha(f)+\alpha^*(f)}{\alpha_0}\right)2eI_c + 4kT\,\Delta G_e. \tag{10.1d}$$

The correlation power spectrum associated with $i_b^* i_c$ is

$$w_{bc} = w_{ec} - w_c = \left(\frac{\alpha(f)}{\alpha_0} - 1\right)2eI_c. \tag{10.1e}$$

We note that, at low frequencies where $\alpha(f) = \alpha_0$,

$$w_e = 2eI_e, \qquad w_c = 2eI_c, \qquad w_b = 2eI_b, \qquad w_{bc} = 0, \tag{10.4}$$

and

$$w_{ec} = 2eI_c.$$

This completes our calculation of the basic noise processes. At low frequencies each current displays full shot noise, and the base and collector currents are uncorrelated. At higher frequencies the base-current noise increases and is partially correlated with the collector-current noise. Because i_b and i_c determine i_e, we are not particularly interested in the behaviour of i_e. The fundamental noise equivalent circuit of the active device is shown in Fig. 10.2 and is valid whatever equivalent circuit we use to represent the properties of the active device.

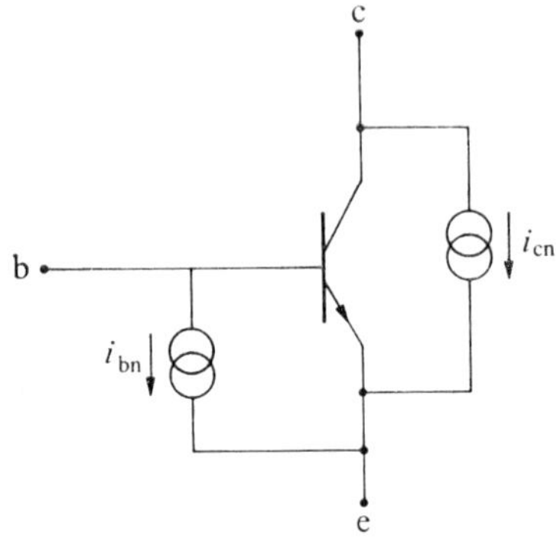

FIG. 10.2. Current noise generators for bipolar transistor.

10.2. Equivalent circuits and parasitic elements

If we neglect the collector–base conductance, which is usually exceedingly small, the signal properties of the basic active device can be represented by

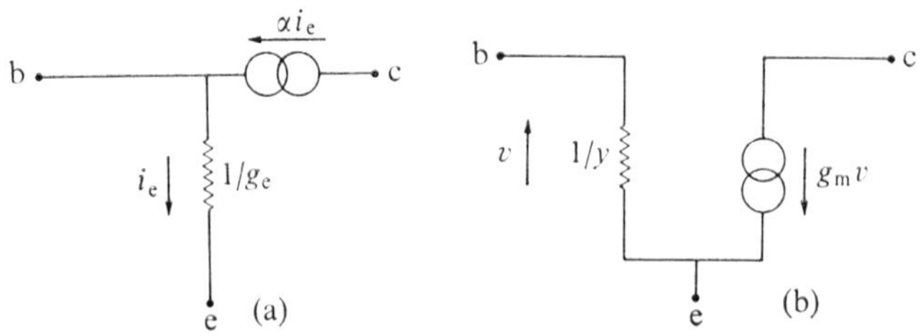

FIG. 10.3. Alternative bipolar transistor equivalent circuits.

either Fig. 10.3(a) or Fig. 10.3(b), where g_e is the differential conductance of the emitter–base diode and

$$g_m = \alpha g_e, \tag{10.3a}$$

$$y = (1-\alpha)g_e. \tag{10.3b}$$

We shall find it useful to have a symbol to denote an idealized device with only the mutual conductance property; this is shown in Fig. 10.4(a). Its equivalent circuit is shown in Fig. 10.4(b), and its relation to a real transistor with all its parasitic elements is shown in Fig. 10.4(c).

The feedback capacitance C_{bc} has a profound effect on the signal and noise properties in high-gain stages, and it cannot be neutralized because of the presence of R_b. It is, however, insignificant in low-gain stages or stages

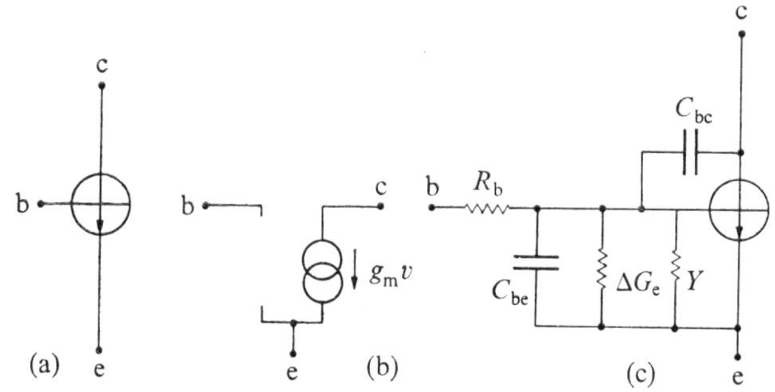

FIG. 10.4(a), (b), and (c). Idealized bipolar transistor symbol, its equivalent circuit (b), and the equivalent circuit (c) of a transistor with its parasitic elements.

driving a second stage of low-input impedance; we shall assume that this is always the case and ignore it.

The input admittance $Y = (1-\alpha)g_e$ is partly real and, at high frequencies, partially imaginary and capacitative, because α is complex. The base–emitter capacitance C_{be} in part represents the effects of electrons making multiple transits and in part is the direct capacitance of the junction. If the multiple transit part is C' it is related to ΔG_e and the mean dwell time τ' of electrons making multiple transits by

$$\Delta G_e = (2\pi f)^2 \tau' c'. \tag{10.4}$$

If we use the familiar approximation

$$\alpha(f) = \frac{\alpha_0}{1+\mathrm{j}f/f_\alpha} \tag{10.5}$$

we have

$$Y = (1-\alpha_0)g_e + \mathrm{j}\frac{f}{f_\alpha}\,g_e = (1-\alpha_0)g_e + \mathrm{j}2\pi f C_\alpha. \tag{10.6a}$$

To this approximation the mutual conductance is $g_m = \alpha_0 g_e \sim g_e$, and the gain bandwidth product is

$$f_T = \frac{g_m}{2\pi(C_\alpha + C_{be})} = \frac{f_\alpha}{1+2\pi f_\alpha C_{be}/g_e}. \tag{10.6b}$$

As the bias current and g_e are increased f_T approaches f_α and, experimentally, there is usually an appreciable range of bias currents for which f_T is almost constant and equal to f_α. Thus in normal operation $C_{be} \ll C_\alpha$ and $2\pi f C_{be}$ is much less than fg_e/f_α. Since C' is only a part of C_{be} it follows that

$$\Delta G_e \ll 2\pi f \tau' g_e (f/f_\alpha).$$

Since we expect τ', the dwell time, to be less than τ the base transit time which determines f_α we have also

$$\Delta G_e \ll g_e (f/f_\alpha)^2.$$

To second order in f/f_α we have

$$Y = \left(1 - \frac{\alpha_0}{1+jf/f_\alpha}\right) g_e = \{1-\alpha_0+j(f/f_\alpha)+(f/f_\alpha)^2\} g_e,$$

and so we can certainly neglect ΔG_e in comparison with Y. Further ΔG_e appears in the noise (see eqn (10.1d)) in conjunction with a term

$$\left(2 - \frac{\alpha+\alpha^*}{\alpha_0}\right) 2eI_c \sim 4eI_c \frac{(f/f_\alpha)^2}{1+(f/f_\alpha)^2} < 4kTg_e(f/f_\alpha)^2,$$

and so the effects of ΔG_e are negligible here as well. Henceforth we shall omit ΔG_e entirely, also ignoring C_{be} and the distinction between f_α and f_T. Because f_T is the parameter most usually quoted in data sheets we shall use it in preference to f_α.

We are now left with R_b as the only important parasitic element. Its effect on the signal response depends on the magnitude of $R_b Y$, i.e. on

$$|R_b Y| = g_e R_{be} |1-\alpha_0+j(f/f_T)| = g_e R_b \{(1-\alpha_0)^2+(f/f_T)^2\}^{\frac{1}{2}}.$$

At low frequencies this is small, if $g_e R_b \ll \beta_0$, where

$$\beta_0 = \frac{\alpha_0}{1-\alpha_0} \approx \frac{1}{1-\alpha_0}, \tag{10.7}$$

is the low-frequency differential current gain, and this is almost always the case. If, however, f approaches f_T we cannot usually neglect R_b and indeed it is a critical parameter in determining the upper frequency limit at which there is power gain. Its effects on the noise performance are, as we shall see, more serious.

The noise current generator i_{cn} in Fig. 10.2 is equivalent to a voltage generator $v_n = i_{cn}/g_m$, as shown in Fig. 10.5(a).

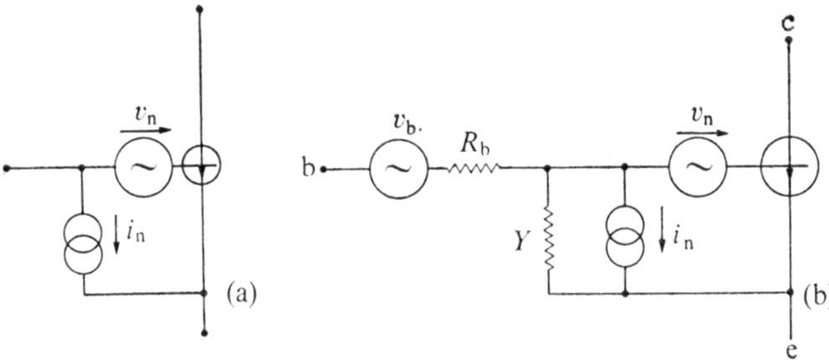

FIG. 10.5(a) and (b). Noise sources and their location for (a) an ideal transistor and (b) for a real transistor.

The power spectra appropriate to this figure are then

$$w_i = 2eI_b + 2eI_c\left(2 - \frac{\alpha(f)+\alpha^*(f)}{\alpha_0}\right), \tag{10.8a}$$

$$w_v = \frac{2eI_c}{g_m(f)g_m^*(f)}, \tag{10.8b}$$

$$w_{iv} = \frac{2eI_c}{g_m}\left(\frac{\alpha(f)}{\alpha_0} - 1\right), \tag{10.8c}$$

and the power spectrum of the Johnson noise associated with R_b, in Fig. 10.5(b), is

$$w_b = 4kTR_b \tag{10.8d}$$

Since the low-frequency mutual conductance is

$$g_m = \alpha_0 g_e = \alpha_0 \frac{eI_e}{kT} = \frac{\alpha_0}{\alpha_{dc}}\frac{eI_c}{kT} \approx \frac{eI_c}{kT} \tag{10.9}$$

and, at higher frequencies,

$$g_m(f) = g_m\alpha(f)/\alpha_0, \tag{10.10}$$

we see that we can also express these results as

$$w_i = \left\{\frac{1}{\beta_{dc}} + \left(2 - \frac{\alpha(f)+\alpha^*(f)}{\alpha_0}\right)\right\} 2kTg_m, \tag{10.11a}$$

$$w_v = \frac{\alpha_0^2}{\alpha(f)\alpha^*(f)} \cdot \frac{2kT}{g_m}, \tag{10.11b}$$

and

$$w_{iv} = 2\left(1 - \frac{\alpha_0}{\alpha(f)}\right)kT, \tag{10.11c}$$

where

$$\beta_{dc} = I_c/I_b = h_{FE} \tag{10.12}$$

is the d.c. current gain. In terms of eqn (10.5) we have

$$w_i = \left\{\frac{1}{\beta_{\mathrm{dc}}} + \frac{2(f/f_T)^2}{1+(f/f_T)^2}\right\} 2kTg_{\mathrm{m}}, \tag{10.13a}$$

$$w_v = \{1+(f/f_T)^2\}\frac{2kT}{g_{\mathrm{m}}}, \tag{10.13b}$$

$$w_{iv} = -2\mathrm{j}(f/f_T)kT. \tag{10.13c}$$

It is often also useful to define an equivalent noise resistance r_{n}, so that $w_v = 4kTr_{\mathrm{n}}$, and an equivalent noise conductance g_{n}, so that $w_i = 4kTg_{\mathrm{n}}$, and to express w_{iv} as $c(w_i w_v)^{\frac{1}{2}}$. Eqns (10.13) yield these quantities as

$$g_{\mathrm{n}} = \frac{1}{2}\left\{\frac{1}{\beta_{\mathrm{dc}}} + \frac{2(f/f_T)^2}{1+(f/f_T)^2}\right\} g_{\mathrm{m}}, \tag{10.14a}$$

$$r_{\mathrm{n}} = \frac{1}{2g_{\mathrm{m}}}\{1+(f/f_T)^2\}, \tag{10.14b}$$

and

$$c = \frac{-\mathrm{j}\theta}{(1/\beta+2\theta^2+\theta^2/\beta)^{\frac{1}{2}}} \approx \frac{-\mathrm{j}\theta}{(1/\beta+2\theta^2)^{\frac{1}{2}}}. \tag{10.14c}$$

we see that $cc^* < \frac{1}{2}$, and c itself is purely imaginary. Thus any correlation between i_{n} and v_{n} is between components which differ in phase by 90°.

The noise due to R_{b} has to be compared with the noise described by $r_{\mathrm{n}} \sim 1/2g_{\mathrm{m}}$. Thus for R_{b} to be negligible we require $R_{\mathrm{b}}g_{\mathrm{m}} \ll \frac{1}{2}$. This can only be achieved by working at very low collector currents and low values of g_{m}. If I_{c} is too small, however, both f_T and β_{dc} are decreased, f_T because we can no longer assume that $C_\alpha \gg C_{\mathrm{be}}$ and β_{dc} because recombination currents in the base–emitter junction, which do not contribute to I_{c}, assume an increased significance. In general we cannot ignore R_{b}.

We have remarked that c is purely imaginary and that $cc^* < \frac{1}{2}$; we shall see in Chapter 12 that this means that the correlation has little effect on either the design or the performance of the low-noise amplifiers. For most purposes we can ignore it, and eqns (10.14a) and (10.14b) together with (10.8d) are an adequate description of the noise. Furthermore, we see from (10.14a) and (10.14b) that if

$$f/f_T \ll (2\beta_{\mathrm{dc}})^{-\frac{1}{2}}. \tag{10.15}$$

we can also ignore the frequency-dependent terms and use

$$w_i = 4kTg_{\mathrm{n}} = 2kTg_{\mathrm{m}}/\beta_{\mathrm{dc}}, \tag{10.16a}$$

$$w_v = 4kTr_{\mathrm{n}} = 2kT/g_{\mathrm{m}}, \tag{10.16b}$$

$$w_{iv} = 0, \tag{10.16c}$$

$$w_{\mathrm{b}} = 4kTR_{\mathrm{b}}. \tag{10.16d}$$

If, for example, $\beta_{dc} = 200$ and $f_T = 400$ MHz, these results will be adequate up to about 20 MHz. (But see also Appendix B.)

We note that, because of the presence of R_b, the collector-current noise, with the base shorted to the emitter, is increased from $w_c = 2eI_c$ to

$$w_c = 2eI_c(1+2g_m R_b) = 2eI_c+4e^2I_c^2R_b/kT. \qquad (10.17)$$

If $R_b = 200\ \Omega$ and $I_c = 1$ mA the increase is sixteen-fold.

10.3. Flicker noise

The average transistor displays a very considerable excess of noise at low frequencies but if, by careful design, all contributions due to surface effects are limited the residual excess noise is due solely to fluctuations in the rate of recombination of minority carriers in the base. This has little effect on the collector-current noise but increases the base-current noise by an amount proportional to the base current and inversely proportional to the frequency. We may write this extra noise as $2eI_b f_0/f$, so that

$$w_i = 2eI_b(1+f_0/f) = 2kTg_m\frac{(1+f_0/f)}{\beta_{dc}} \qquad (10.18)$$

giving

$$g_n = (1+f_0/f)\frac{g_m}{2\beta_{dc}}. \qquad (10.19)$$

The constant f_0 varies from device to device but is generally of the order of 1 kHz.

10.4. Summary

At moderate frequencies transistor noise is described basically in terms of shot noise in the base and collector currents and Johnson noise in the base resistance. At low frequencies the base–current noise increases, and it also increases at high frequencies. Of the five significant parameters g_m, β_{dc}, f_T, R_b, and f_0 the first three are readily available from device data sheets; however, the values of R_b and f_0 are quoted less frequently and, at best, have to be estimated from other data. This is unfortunate as they both exert a significant influence on the noise performance. Some manufacturers provide data on the equivalent r.m.s. input noise fluctuations with the base connected to the emitter. If this is denoted by e_n we have, at low frequencies,

$$e_n^2 = 4kTR_b+2R_b^2eI_b(1+f_0/f)+2kT/g_m,$$

and so, if e_n is given as a function of I_b (or I_c/β_{dc}) and f, it is possible to deduce the values of R_b and f_0.

11 Field-Effect Transistors

11.1. Introduction

A field-effect transistor (f.e.t.) is essentially a capacitor in which one plate, the channel, consists of a thin layer of semiconductor. If this is n-type material, a negative potential applied to the other plate, the gate G, requires a net positive charge in the channel, and so results in a reduced carrier concentration. This modulates the conductance of the channel and can be used to control a current flowing between two contacts: the source S and the drain D at the ends of the channel. The exact details of the operation depend on the particular structure, but the simple model which we shall adopt describes adequately the behaviour of most devices, and indeed its predictions are in somewhat better accord with experiment than many more-sophisticated models.

Fig. 11.1 indicates conventions for the current and voltages with a channel of over-all length L_0. If the mobile charge per unit length of the channel is $-q$ and the mobility of the carriers is $-\mu$, the resistance per unit length is $1/\mu q$, and so if ϕ is the channel potential at x measured relative to the gate,

$$I = -\mu q \frac{\partial \phi}{\partial x}. \tag{11.1}$$

Since I is independent of x this leads to the relation

$$I(x_2 - x_1) = -\int_{\phi_1}^{\phi_2} \mu q \, d\phi = -\int_{\phi_1}^{\phi_2} g \, d\phi, \tag{11.2}$$

where

$$g = \mu q. \tag{11.3}$$

If this result is applied to the source at $x_1 = 0$ and the drain at $x_2 = L_0$ we have for the source and drain conductances,

$$G_s = \frac{\partial I}{\partial V_s} = \frac{\partial I}{\partial \phi_s} = \frac{g_s}{L_0}, \tag{11.4a}$$

$$G_d = -\frac{\partial I}{\partial V_d} = -\frac{\partial I}{\partial \phi_d} = \frac{g_d}{L_0}, \tag{11.4b}$$

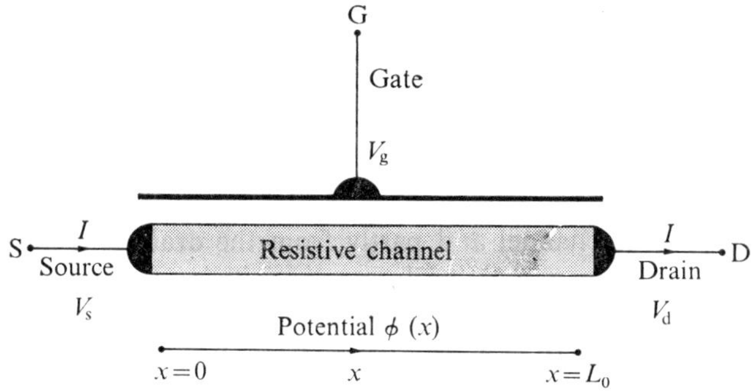

FIG. 11.1. A field-effect transistor.

and the mutual conductance is

$$G_{\mathrm{m}} = -\frac{\partial I}{\partial V_{\mathrm{g}}} = \frac{\partial I}{\partial \phi_{\mathrm{s}}} + \frac{\partial I}{\partial \phi_{\mathrm{d}}} = G_{\mathrm{s}} - G_{\mathrm{d}} = \frac{g_{\mathrm{s}} - g_{\mathrm{d}}}{L_0}. \tag{11.4c}$$

The mobile charge q per unit length is related to the mobile charge q_0 in the unperturbed channel by

$$-q = -q_0 + \gamma\phi,$$

where γ is the capacitance per unit length between gate and channel. Since the mobility $-\mu$ can be regarded as constant we have

$$g = \mu q = \mu q_0\left(1 - \frac{\gamma\phi}{q_0}\right) = g_0\left(1 - \frac{\phi}{V_{\mathrm{p}}}\right), \tag{11.5}$$

where

$$V_{\mathrm{p}} = q_0/\gamma$$

is the voltage required to deplete the channel completely, i.e. the pinch-off voltage. In terms of g_0 and V_{p} eqn (11.1) becomes

$$I = -g_0\left(1 - \frac{\phi}{V_{\mathrm{p}}}\right)\frac{\partial \phi}{\partial x}, \tag{11.7}$$

which can be integrated immediately to give

$$I = -\frac{g_0 V_{\mathrm{p}}}{L_0}\left(\frac{\phi_{\mathrm{d}} - \phi_{\mathrm{s}}}{V_{\mathrm{p}}} - \frac{1}{2}\,\frac{\phi_{\mathrm{d}}^2 - \phi_{\mathrm{s}}^2}{V_{\mathrm{p}}^2}\right). \tag{11.8}$$

For small values of ϕ_{d} and ϕ_{s} this reduces to

$$I = -G_0(V_{\mathrm{d}} - V_{\mathrm{s}}), \tag{11.9}$$

where

$$G_0 = \frac{g_0}{L_0} = \frac{\mu q_0}{L_0} \tag{11.10}$$

is the conductance of the unbiased channel.

The potential ϕ of the channel relative to the gate increases towards the drain and, in the usual mode of operation, considerably exceeds V_p near the drain. The channel is not however, as indicated by our analysis so far, completely pinched-off unless ϕ_s also exceeds V_p. Instead in a region near the drain the channel is almost depleted of mobile charge. In this region, lines of force, terminating on the negative gate, originate partly from unbalanced positive ions in the channel and partly from the drain contact itself. The situation is illustrated in Fig. 11.2. The conductivity in the pinched region near the drain is low, and most of the potential difference between drain and source occurs across this region.

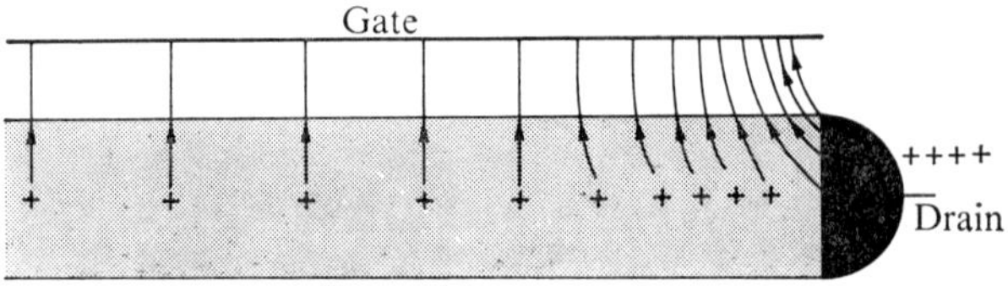

FIG. 11.2. Charge and lines of force near the drain in a pinched channel. Near the drain lines of force originate not only from the un-neutralized donor ions in the channel but also from the drain contact itself.

Because g near the drain is small, G_d is also small and can be neglected. The mutual conductance G_m is then equal to G_s. The pinched region occupies an appreciable, but not large, fraction of the length L_0 of the channel, but only the region between $x = 0$ and $x = L$, where $\phi = V_p$, determines the signal and noise properties of the device. In this region eqn (11.7) is valid, and I is given by (11.8) with V_d replaced by V_p and L_0 by L. This yields

$$I = -\frac{g_0 V_p}{2L}\left(1-\frac{\phi_s}{V_p}\right)^2. \tag{11.11}$$

If we now take $V_s = 0$, then I is related to the gate voltage V_g, which is negative, by

$$I = -\frac{g_0 V_p}{2L}\left(1+\frac{V_g}{V_p}\right)^2 = I_0\left(1+\frac{V_g}{V_p}\right)^2, \tag{11.12}$$

where I_0 is the current at zero gate bias with $V_d > V_p$. Eqn (11.12) is the basic f.e.t. equation and is in good agreement with the observed properties of the device. The mutual conductance obtained from (11.12) is

$$G_m = -2\frac{I_0}{V_p}\left(1+\frac{V_g}{V_p}\right) = \frac{g_0}{L}\left(1+\frac{V_g}{V_p}\right), \tag{11.13}$$

and, at zero bias, $G_m = G_0$, the unbiased channel conductance.

The total channel charge, which responds to changes in bias, is the integral of $-q$ from the source to the point L where $\phi = V_p$. Beyond this point there is very little mobile charge in the channel. The current I is carried by what little charge remains, and the charge moves at a high velocity because the potential gradient is large in this region. The variable gate-charge is therefore

$$Q = \int_0^L q \, dx = q_0 \int_0^L (1-\phi/V_p) \, dx,$$

which can also be expressed as

$$Q = \frac{q_0}{g_0} \int_0^L g \, dx. \tag{11.14}$$

Integrals of the form

$$H_n = \int_0^L g^n \, dx \tag{11.15}$$

will appear frequently in our calculations. They may be evaluated as follows.

$$H_n = \int_{\phi_s}^{V_p} g^n \frac{dx}{d\phi} \, d\phi = -\frac{1}{I} \int_{\phi_s}^{V_p} g^{n+1} \, d\phi = -\frac{g_0^{n+1} V_p}{I} \int_{\phi_s}^{V_p} \left(1 - \frac{\phi}{V_p}\right)^{n+1} \frac{d\phi}{V_p}$$

and, since

$$I = \frac{g_0 V_p}{2L} \left(1 - \frac{\phi_s}{V_p}\right)^2,$$

this gives

$$H_n = \frac{2}{n+2} g_0^n L \left(1 - \frac{\phi_s}{V_p}\right)^n. \tag{11.16}$$

We have therefore

$$Q = \frac{q_0}{g_0} H_1 = \tfrac{2}{3} q_0 L \left(1 + \frac{V_g - V_s}{V_p}\right),$$

and so the input capacitance is

$$C = \frac{\partial Q}{\partial V_g} = \frac{2}{3} \frac{q_0 L}{V_p} = \tfrac{2}{3} \gamma L. \tag{11.17}$$

We see that the gain–bandwidth product is

$$f_T = \frac{G_m}{2\pi C} = \frac{3}{4\pi} \frac{V_p G_0}{q_0 L} \left(1 + \frac{V_g}{V_p}\right) = \frac{3}{4\pi} \frac{\mu V_p}{L^2} \left(1 + \frac{V_g}{V_p}\right). \tag{11.18}$$

This is a maximum at zero bias and, if f_T is to be large, the channel must be short and the carriers have a high mobility.

The input capacitance C is charged by current flowing in the channel and so in an equivalent circuit appears in series with a part of the channel resistance of the order of $1/G_m$. The equivalent circuit of the device, whose conventional

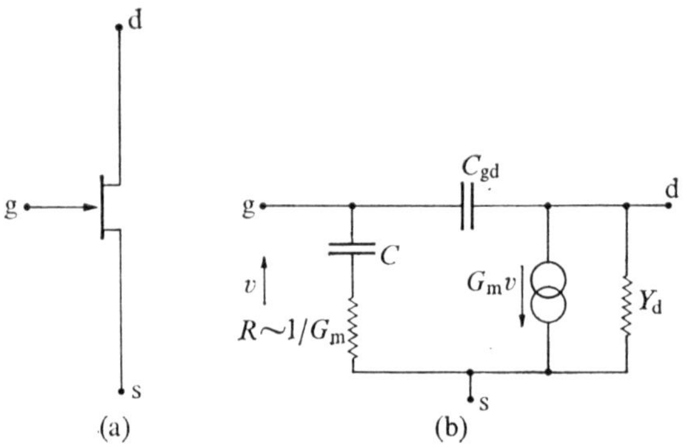

FIG. 11.3(a) and (b). Circuit symbol and equivalent circuit for an f.e.t.

circuit symbol is shown in Fig. 11.3(a), is shown in Fig. 11.3(b). The parasitic gate–drain capacitance C_{gd} and the small, but finite, output admittance Y_d are also indicated.

11.2. Noise

Electrons in the active region between the source and $x = L$, where $\phi = V_p$, are in thermal equilibrium with the lattice of the channel and so generate Johnson noise. The power spectrum of the voltage fluctuations v_1, generated across a length dx_1 at x_1, is $4kT\,dx_1/g_1$. Electrons in the pinched region between L and L_0 are not in thermal equilibrium, but nevertheless their statistical properties are determined by their very frequent collisions with lattice atoms. They can be included in the same way if we allow T_1 to be a function of position x_1.

If all the electrodes are shorted (for a.c.), a fluctuation v_1 generated across dx_1 at x_1 leads to a change i_1 in the channel current and a charge $\delta_1\phi(x)$ in the channel potential at all points between the source and drain. If eqn (11.2) is applied first to the region between the source and x_1 and then to the region between x_1+dx_1 and the drain, we obtain

$$i_1 = -\frac{g_1}{x_1}\,\delta_1\phi_1 \tag{11.19a}$$

$$i_1 = \frac{g_1}{L_0-x_1}(\delta_1\phi_1+v_1), \tag{11.19b}$$

where $\delta_1\phi_1$ is the change in channel potential at x_1. We can eliminate $\delta_1\phi_1$ and obtain

$$i_1 = \frac{g_1}{L_0}v_1. \tag{11.20}$$

Now fluctuations such as v_1 generated at different points x_1 are uncorrelated, and so the power spectrum of the total channel current is simply the integral

$$w_{ic} = \int_0^{L_0} \left(\frac{g(x)}{L_0}\right)^2 \frac{4kT(x)}{g(x)}\,dx = \frac{1}{L_0^2}\int_0^{L_0} 4kT(x)g(x)\,dx. \tag{11.21}$$

Between $x = 0$ and $x = L$ we have $T(x) = T$, the channel temperature, but beyond $x = L$ up to the drain at $x = L_0$ we have $T(x) > T$. On the other hand, $g(x)$ is large for $x < L$ and exceedingly small for $x > L$. Thus, unless $T(x)$ is extremely large in this region, which generally only occurs just before breakdown, we can neglect contributions to w_{ic} from the pinched region and replace L_0 in (11.21) by L and $T(x)$ by T. This gives

$$w_{ic} = \frac{4kT}{L_0^2}\int_0^L g\,\mathrm{d}x = \frac{4kTH_1}{L_0^2} = \frac{4kTg_0}{L}\cdot\frac{2}{3}\left(1+\frac{V_g-V_s}{V_p}\right),$$

which can be expressed also in terms of the mutual conductance G_m as

$$w_{ic} = \tfrac{2}{3}.4kTG_m. \tag{11.22}$$

Fluctuations in the channel potential lead to fluctuations in the gate charge. From eqns (11.2), (11.19a), and (11.19b) we obtain the change in channel potential at a general point x due to a fluctuation v_1 across $\mathrm{d}x_1$ at x_1 as

$$\delta_1\phi = -\frac{xg(x_1)v_1}{Lg(x)}, \qquad x < x_1 \tag{11.23a}$$

$$\delta_1\phi = \frac{(L-x)g(x_1)v_1}{Lg(x)}, \qquad x > x_1 \tag{11.23b}$$

and so the resulting fluctuation in the gate charge Q is, from (11.14),

$$\delta_1 Q = \frac{q_0}{g_0}\int_0^L \frac{\mathrm{d}g}{\mathrm{d}\phi}\delta_1\phi\,\mathrm{d}x, \tag{11.24}$$

which gives

$$\delta_1 Q = \frac{q_0}{g_0}\left(\int_{x_1}^L \frac{1}{g}\frac{\mathrm{d}g}{\mathrm{d}\phi}\,\mathrm{d}x - \frac{1}{L}\int_0^L \frac{x}{g}\frac{\mathrm{d}g}{\mathrm{d}\phi}\,\mathrm{d}x\right)g(x_1)v_1. \tag{11.25}$$

Since $\mathrm{d}x/g = -\mathrm{d}\phi/I$ this yields

$$\delta_1 Q = \frac{q_0 g(x_1)v_1}{g_0 I}\left(g(x_1) - \frac{1}{L}\int_0^L g\,\mathrm{d}x\right) = \frac{q_0 g(x_1)v_1}{g_0 I}\left(g(x_1) - \frac{H_1}{L}\right). \tag{11.26}$$

The noise current flowing *into* the gate is $i_g = j\omega Q$, and so the induced gate noise-current power spectrum is

$$w_{ig} = \omega^2\left(\frac{q_0}{g_0 I}\right)^2 \int_0^L g(x_1)\left(g(x_1) - \frac{H_1}{L}\right)^2 4kT\,dx_1. \tag{11.27}$$

This involves the integrals of g, g^2, and g^3, and we obtain

$$w_{ig} = \omega^2\left(\frac{q_0}{g_0 I}\right)^2 4kT\left(H_3 - 2\frac{H_1 H_2}{L} + \frac{H_1^3}{L^2}\right) = \frac{4}{15}\cdot\frac{\omega^2 C^2}{G_m}.4kT. \tag{11.28}$$

Because the gate current fluctuations arise from the same fundamental random process they are correlated with the channel-current fluctuations. The correlation can be obtained using eqns (11.20) and (11.26), and with $i_g^* = -j\omega Q$ we obtain

$$w_{gc} = -j\omega\frac{q_0 . 4kT}{g_0 I}\int_0^L g(x)\left\{g(x) - \frac{H_1}{L}\right\} dx = \frac{4j\omega CkT}{6}. \tag{11.29}$$

As a measure of the extent of the correlation we have

$$c^2 = \frac{w_{gc} w_{gc}^*}{w_{ic} w_{ig}} = \frac{5}{32} \sim 0{\cdot}16. \tag{11.30}$$

In junction transistors the gate leakage current I_g adds a further uncorrelated component to the gate-current fluctuations, i.e.

$$w_1 = 2eI_g, \tag{11.31}$$

and we have

$$\frac{w_{ig}}{w_1} = \frac{8\omega^2 C^2}{15 G_m}\frac{kT}{eI_g}. \tag{11.32}$$

Below a frequency

$$f \sim \frac{1}{2\pi C}\left(\frac{15eI_g G_m}{8kT}\right)^{\frac{1}{2}} \sim f_T\left(\frac{2eI_g}{G_m kT}\right)^{\frac{1}{2}} \sim f_T\left(\frac{80 I_g}{G_m}\right)^{\frac{1}{2}}, \tag{11.33}$$

the leakage-current noise is larger; this frequency is usually of the order of 100 kHz to 1 MHZ.

In most devices the gate does not quite cover the whole channel and the inactive region near the source represents a small series resistance, usually much less than $1/G_0$, in the source lead. Its effects are, however, generally insignificant.

11.3. Equivalent circuits

In terms of the gate and collector currents the noise equivalent circuit is Fig. 11.4(a), but we can also transfer i_c to the gate circuit as a voltage generator $v_n = -i_c/G_m$. This is shown in Fig. 11.4(b), and the relevant power

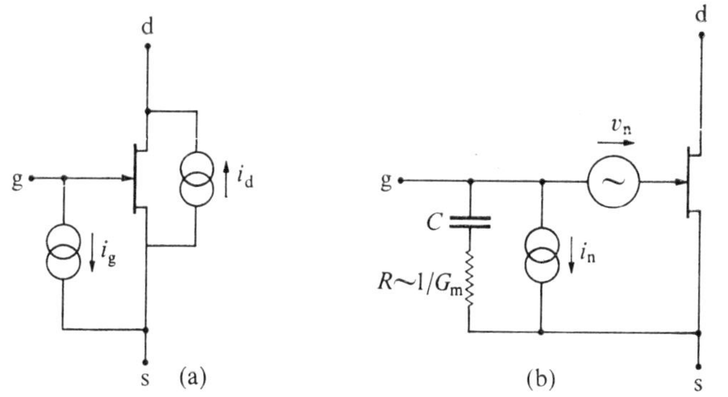

FIG. 11.4(a) and (b). Noise in an f.e.t. represented in terms of either two current generators or one current generator and one voltage generator.

spectra are

$$w_v = \frac{2}{3} \cdot \frac{4kT}{G_m}, \tag{11.34a}$$

$$w_i = \frac{4}{15} \cdot \frac{\omega^2 C^2}{G_m} . 4kT + 2eI_g, \tag{11.34b}$$

$$w_{iv} = -\frac{1}{6} j\omega C . \frac{4kT}{G_m}. \tag{11.34c}$$

The behaviour of an f.e.t. is very similar to that of a triode. If we note that the real part of the input admittance is approximately $G_{in} \sim \omega^2 C^2/G_m$ then, at high frequencies, $w_i \sim \frac{1}{4} G_{in} 4kT$, whereas with a triode $w_v \sim 2{\cdot}5\,(4kT/G_m)$ and $w_i \sim 5G_{in}4kT$.

11.4. Dual-gate devices

The dual-gate f.e.t., with its reduced Miller capacitance, is the analogue of the vacuum tetrode or pentode. There is, however, no equivalent to partition noise, and so the improved signal performance is achieved without any increase in the noise.

11.5. Flicker noise

In devices not specially designed for low-noise low-frequency amplifiers there is a very considerable increase in the noise output at low frequencies, and this is associated mainly with surface phenomena. These surface effects can be eliminated, however, in junction-gate devices, and the residual noise is then due to fluctuations in the recombination rate of carriers in the gate–channel depletion layer. These fluctuations modulate the channel current and therefore are equivalent to an additional noise voltage generator in the gate circuit. There is little effect on the gate-current noise, and so at low frequencies the noise equivalent circuit has the form shown in Fig. 11.5. The

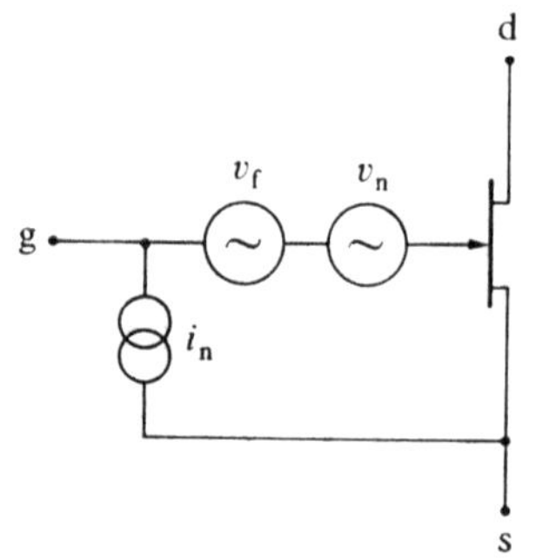

FIG. 11.5. The extra flicker-noise generator v_f for f.e.t. transistor.

power spectra of the two white noise generators are

$$w_i = 2eI_g \tag{11.35a}$$

and

$$w_v = \frac{2}{3}.\frac{4kT}{G_m}. \tag{11.35b}$$

It is convenient to express the flicker-noise power spectrum as

$$w_f = w_v\frac{f_0}{f} = \frac{2}{3}.\frac{4kT}{G_m}.\frac{f_0}{f}, \tag{11.35c}$$

and the frequency f_0 determines frequency at which flicker noise first becomes noticeable. For many low-noise devices this is about 20–30 kHz.

11.6. Insulated-gate devices

At moderate to high frequencies there is little difference between junction- and insulated-gate devices, but at low frequencies the much higher flicker noise of insulated-gate devices makes them totally unsuitable for low-noise amplifiers.

12 Linear Amplifiers

12.1. Introduction

The signal response of an amplifier with a definite output load may be described in terms of its frequency-dependent input admittance $Y(f)$ and gain $A(f)$. Its noise performance is specified by the power spectrum $W(f)$ of its output when it is connected to a definite source. Although all the fundamental sources of noise have a white or, at least, slowly varying spectrum, the spectrum $W(f)$ is modified by the frequency response of the amplifier, and the mean square output fluctuations

$$\langle\Delta\theta\rangle^2 = \int_0^\infty W(f)\,\mathrm{d}f \tag{12.1}$$

will be finite. If θ_s is the r.m.s. value of the output signal $\theta_s(t)$, the signal to noise ratio at the output is given by

$$\left(\frac{S}{N}\right)^2 = \frac{\theta_s^2}{\langle\Delta\theta\rangle^2}, \tag{12.2}$$

and this describes the over-all performance of the amplifier for a specified form of the signal input. It is not, however, a very convenient description, and we shall find it more useful to consider single Fourier components of the signal and the noise in a small bandwidth $\mathrm{d}f$ centred on the signal frequency. Thus, if $\theta(f)$ is the r.m.s. signal output at f, we shall consider the ratio of $\theta^2(f)$ to $W(f)\,\mathrm{d}f$.

The output-noise power derives from sources of more or less white spectral density, and we shall be particularly interested in how much of $W(f)$ arises from internal sources and how much arises from noise present in the signal input. This input noise will almost always be white. Thus it is useful to express $w(f)$ in the form

$$W(f) = A^*(f)A(f)\,w \tag{12.3}$$

in terms of the white noise w at the input which would give the same output noise, in an otherwise noiseless system of the same gain $A(f)$.

The gain $A(f)$ may either rise to a peak at some frequency f_0 and fall off rapidly at frequencies different from f_0 or it may be more or less constant over a range of frequencies including f_0 and fall to zero at other frequencies. Usually signals will be confined to frequencies where $A(f)$ has its maximum value or a more or less constant value, which we denote by A_0. It is then convenient to define an effective bandwidth Δf by

$$\Delta f = \int_0^\infty \frac{A^*(f)A(f)\,\mathrm{d}f}{A_0^*A_0}, \tag{12.4}$$

so that the output fluctuations can be expressed as

$$\langle\Delta\theta^2\rangle = \int_0^\infty W(f)\,\mathrm{d}f = wA_0^*A_0\,\Delta f. \tag{12.5}$$

The mean square value s^2 of a signal within the pass-band (i.e. at a frequency where $A(f) = A_0$) that gives unity signal to noise ratio is then

$$s^2 = w\,\Delta f. \tag{12.6}$$

For most purposes w and Δf define the sensitivity of the system.

If the amplifier consists of a cascade of n stages with gains $A_1(f)$, $A_2(f)$, etc., and the power spectra of the noise voltages introduced at the inputs to each stage are w_1, w_2, etc., we have

$$W = w_1A_1^*A_1A_2^*A_2...A_n^*A_n + w_2A_2^*A_2...A_n^*A_n + ... + w_nA_n^*A_n, \tag{12.7}$$

and most of the output noise arises from the earlier stages. The equivalent input noise is

$$w = w_1 + w_2/A_1^*A_1 + ... + w_n/A_1^*A_1...A_{n-1}^*A_{n-1}. \tag{12.8}$$

However, the over-all response and especially the bandwidth is determined by all the stages. Since the signal has to traverse every stage, the bandwidth of each stage must accommodate the signal, but it should not be unnecessarily large, because every increment of bandwidth allows additional noise components to reach the output. Excessive bandwidth in the earlier stages does not usually matter provided that at least the last stage has the correct bandwidth. It will be convenient to assume that the over-all bandwidth of the system is always set by the last few stages, and that these occur at a level where the amplified noise from the earlier stages is large enough to make noise generated in these last few stages negligible. We can then assume that the amplifier has a well-defined frequency response and effective bandwidth, without having to discuss bandwidth limitations in the earlier stages. Noise in these stages can be regarded as effectively white.

The performance of the complete system is described by $w\,\Delta f$; w is determined mainly by the first few stages and Δf by the last few stages or, perhaps, by a filter network in the output.

The effective bandwidth Δf defined by (12.4) is not the same as the usual signal bandwidth $\Delta f_{\frac{1}{2}}$ defined in terms of the half-power points. Thus, for a tuned circuit or Lorentzian response with

$$A(f) = \frac{A_0}{1+\mathrm{j}Q(f/f_0-f_0/f)}, \tag{12.9a}$$

we have

$$\Delta f = \frac{\pi}{2}\frac{f_0}{Q} = \frac{\pi}{2}\Delta f_{\frac{1}{2}}, \tag{12.9b}$$

while for a Gaussian response

$$A^*(f)A(f) = A_0^2 \exp\left\{-\frac{1}{2}\left(\frac{f-f_0}{\Delta f_0}\right)^2\right\} \tag{12.10a}$$

we have

$$\Delta f = 1{\cdot}34\ \Delta f_{\frac{1}{2}}. \tag{12.10b}$$

12.2. Noise figure and noise temperature

If a signal of r.m.s. amplitude $S(f)$ at a definite frequency f is accompanied by noise of power spectrum w_s at f the signal to noise ratio associated with an infinitesimal bandwidth df about f is

$$\left(\frac{S}{N}\right)_i^2 = \frac{s^2(f)}{w_s\, df}. \tag{12.11}$$

If this signal is applied to an amplifier of gain $A(f)$ and output-noise power spectrum $W(f)$ the output signal to noise ratio associated with df is

$$\left(\frac{S}{N}\right)_o^2 = \frac{A^*(f)A(f)s^2(f)}{W(f)\,\mathrm{d}f}. \tag{12.12}$$

The spot frequency noise figure $F(f)$ is then defined as the ratio

$$F(f) = \frac{(S/N)_i^2}{(S/N)_o^2} = \frac{W(f)}{w_s A^*(f)A(f)}. \tag{12.13}$$

The output noise is related to the equivalent input noise w by (12.3) and $w = w_s + w_a$, where w_a is the contribution from the amplifier. We have therefore

$$F(f) = \frac{w}{w_s} = \frac{w_s+w_a}{w_s} = 1+\frac{w_a}{w_s} \geqslant 1. \tag{12.14}$$

The total output fluctuations are

$$\langle \Delta\theta^2 \rangle = \int_0^\infty F(f)\, w_s A^*(f)A(f)\,\mathrm{d}f, \tag{12.15}$$

and, if as is usually the case, w_s is independent of f, or at least varies very little over the pass-band of the amplifier, it is useful to define an average noise figure.

$$\bar{F} = \frac{\int F(f)A^*(f)A(f)\,\mathrm{d}f}{\int A^*(f)A(f)\,\mathrm{d}f} = \frac{\int F(f)A^*(f)A(f)\,\mathrm{d}f}{A_0^*A_0\,\Delta f} \tag{12.16}$$

so that

$$\langle\Delta\theta^2\rangle = \bar{F}A_0^*A_0 w_s\,\Delta f. \tag{12.17}$$

The r.m.s. signal amplitude yielding unity signal to noise ratio at f is then given by

$$s^2(f) = \frac{A_0^*A_0}{A^*(f)A(f)}\bar{F}w_s\,\Delta f. \tag{12.18}$$

If additional white noise of power spectrum w_n is added at the input, the output fluctuations are given by

$$\langle\Delta\theta^2\rangle = \int_0^\infty (w_s F(f)+w_n)A^*(f)A(f)\,\mathrm{d}f = (\bar{F}+w_n/w_s)\int w_s A^*(f)A(f)\,\mathrm{d}f. \tag{12.19}$$

Thus if w_s is known and w_n is a calibrated source the average noise figure $\bar{F}$ can be obtained from the ratio w_n/w_s required to double the output noise.

In some cases, especially if flicker noise is important and the amplifier bandwidth is large, the distinction between the average noise figure $\bar{F}$ and the spot noise figure $F(f)$ is important, but if $F(f)$ is known we can always obtain $\bar{F}$. Henceforth we shall consider only $F(f)$.

The mean square open-circuit output voltage of a signal source of available power P_s and internal impedance $Z = R+\mathrm{j}X$ is

$$V_s^2 = 4P_sR, \tag{12.20a}$$

and if it also generates an available noise power of spectrum w_s the mean square noise voltage in $\mathrm{d}f$ is

$$V_n^2 = 4w_sR\,\mathrm{d}f. \tag{12.20b}$$

The ratio

$$\frac{V_s^2}{V_n^2} = \frac{P_s}{w_s\,\mathrm{d}f} \tag{12.21}$$

is independent of Z and therefore of any lossless transformation of Z. In particular, if the source is a resistive network, a long attenuating transmission line, or transmission path, and we are dealing with Johnson noise at an effective temperature T_s, then $w_s = kT_s$ and

$$\frac{V_s^2}{V_n^2} = \frac{P_s}{kT_s\,\mathrm{d}f}. \tag{12.22}$$

This is the signal to noise ratio of the source. If this source is connected to an amplifier of noise figure F the output signal to noise ratio is less by a factor $1/F$, and we can express this as

$$\left(\frac{S}{N}\right)_o^2 = \frac{P_s}{FkT_s\,df} = \frac{P_s}{k(T_s+T_n)\,df},$$

where

$$T_n = (F-1)T_s, \tag{12.23}$$

represents the additional noise due to the amplifier. It is known as the noise temperature of the amplifier. The equivalent input noise $w = w_s + w_a$ can be expressed in terms of available power rather than voltage, and we then have

$$w = kT_s + w_a = k(T_s + T_n). \tag{12.24}$$

The input noise power w_a due to the amplifier depends on the impedance of the source, thus T_n is not solely a characteristic of the amplifier. However, for some value of the source impedance Z, w_a, and T_n will have a minimum value, and F will also be a minimum. These minimum quantities are solely characteristic of the amplifier and are what we usually mean by the noise temperature and noise figure of an amplifier. If T_{n0} is the minimum noise temperature of an amplifier of effective bandwidth Δf, a signal power

$$P_s = k(T_s + T_{n0})\,\Delta f, \tag{12.25}$$

from a source at an effective temperature T_s will yield unity signal to noise ratio in the output. It is important to realize that T_n depends on the source impedance. An amplifier with $T_{n0} = 10$ kelvins with a source of impedance $10^6\ \Omega$ may well have $T_n > 10^5$ kelvins if the source impedance is $10\ \Omega$.

In practice, noise figures are usually specified relative to a source at 293 K (20 °C), and so

$$T_n = (F-1)293\ \text{K}. \tag{12.26}$$

For example, if $F = 2$ (3 dB), $T_n = 293$ K and the noise figure relative to a source at 1 K is 294 (25 dB). For a more extended discussion of noise figures the reader should consult the original paper by Friis (1944).

12.3. Noise in a single-stage amplifier without feedback

Fig. 12.1 represents a signal source connected to the input of a device such as a vacuum tube, an f.e.t., or a bipolar transistor. All the internal sources of noise in the device are represented by the two generators v and i. The admittance $(G+jB)$ represents both the admittance of the device and the associated coupling circuits including any susceptance associated with the source. No noise is associated with G, for all circuit loss is described by the transformed source resistance R_s and all device noise is described by v and i. The thermal noise associated with R_s is represented by v_t with a power spectrum $4kTR_s$, and the mean square value of the signal-voltage generator v_s

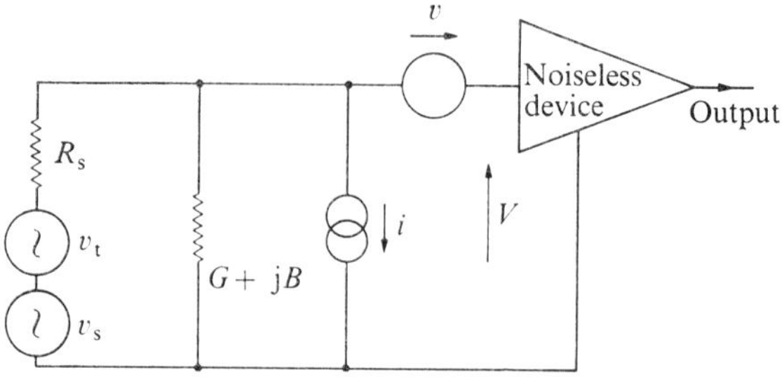

FIG. 12.1. Representation of noise in an amplifier.

is related to the available signal power P_s by $v_s^2 = 4P_sR_s$. If w_i, w_v, and w_{iv} are the power spectra and correlation power spectrum associated with i and v, the power spectrum of the voltage V at the input to the noiseless device is

$$W = w_v + \frac{4kT}{R_s}\{(G+(1/R_s))^2+B^2\}^{-1} + w_i\ (G+(1/R_s))^2+B^2\}^{-1} -$$

$$-w_{iv}\{G+(1/R_s)-jB\}^{-1} - w_{iv}^*\{G+(1/R_s)+jB\}^{-1}, \tag{12.27}$$

while the mean square signal voltage at the same point is

$$V_s^2 = \frac{4P_s}{R_s}\{(G+(1/R_s))^2+B^2\}^{-1}. \tag{12.28}$$

The deterioration in the signal to noise ratio, i.e. the noise factor, is obtained by dividing W by its first term, which is the contribution from the source noise because both this term and V_s^2 are related to kT and P_s by the same factor. This gives

$$4kT_n = 4kT(F-1) = R_s w_i + \{(G+(1/R_s))^2+B^2\}R_s w_v -$$

$$-w_{iv}(G+(1/R_s)+jB)R_s - w_{iv}^*(G+(1/R_s)-jB)R_s. \tag{12.29}$$

The circuit parameters B and R_s can be varied by altering the coupling circuit and, for some set of values, T_n will be a minimum. If we write $w_{iv} = \rho + j\theta$ we see that B occurs only in the combination $B^2 w_v + 2B\theta$, and so the optimum value of B is

$$B = -\frac{\theta}{w_v} = \frac{j(w_{iv}-w_{iv}^*)}{2w_v}. \tag{12.30}$$

With this value of B

$$4kT_n = R_s\left\{w_i + G^2 w_v - \frac{\theta^2}{w_v} - 2\rho G\right\} + 2(Gw_v - \rho) + \frac{w_\sigma}{R_s}. \tag{12.31}$$

This is minimized if R_s is chosen to be equal to

$$R_{s0} = \left\{\frac{w_v}{w_i+G^2w_v-(\theta^2/w_v)-2\rho G}\right\}^{\frac{1}{2}} \tag{12.32}$$

and then T_n has the minimum value T_{n0} given by

$$4kT_{n0} = 2(Gw_v-\rho)+2[w_v\{w_i+G^2w_v-(\theta^2/w_v)-2\rho G\}]^{\frac{1}{2}}. \tag{12.33}$$

In all the devices we have considered w_{iv} is purely imaginary and $\rho = 0$, so that $\theta^2 = w_{iv}w_{iv}^*$ and can be expressed as $c^2w_iw_v$, where

$$c^2 = \frac{w_{iv}^*w_{iv}}{w_iw_v}. \tag{12.34}$$

We then obtain

$$R_{s0} = \left\{\frac{w_v}{(1-c^2)w_i+G^2w_v}\right\}^{\frac{1}{2}}, \tag{12.35}$$

and

$$4kT_{n0} = 2Gw_v+2[w_v\{(1-c^2)w_i+G^2w_v\}]^{\frac{1}{2}}. \tag{12.36}$$

We note that when $R_s \neq R_{s0}$ eqn (12.31) can be written as

$$4kT_n-2Gw_v = \frac{1}{2}\left(\frac{R_s}{R_{s0}}+\frac{R_{s0}}{R_s}\right)(4kT_{n0}-2Gw_v), \tag{12.37}$$

and so the value of T_n is relatively insensitive to small departures of R_s from R_{s0}. Now the maximum value of c^2 that we have encountered is $\frac{1}{2}$ (for a bipolar transistor with zero base-resistance at high frequencies). If we omit c^2 in calculating R_{s0} the error is at most a factor $1/\sqrt{2}$, and this only increases $\frac{1}{2}[(R_s/R_{s0})+(R_{s0}/R_s)]$ from 1 to 1·07. Thus we can certainly ignore correlations in calculating R_{s0}. If we ignore correlations altogether and set $B = 0$ then T_n (see eqn (12.36)) with c^2 set equal to zero is increased by at most $\sqrt{2}$. In practice this is almost negligible but, in any case, we now have a practical design procedure, for we can first calculate R_{s0} neglecting correlations and then, with this value of R_s, adjust B to minimize the noise. If correlations are significant, changing B from zero at most will decrease T_n by $1/\sqrt{2}$. In practice, either because, as in vacuum tubes and f.e.t.s, c^2 is less than $\frac{1}{2}$ or because in bipolar transistors the base resistance reduces the effective value of c^2, the improvement will be much less. Henceforth we shall ignore correlations, and this will greatly simplify our calculations since we can then always deal with purely resistive terms in the input network. We now have $4kT_n$ given by (12.37) with

$$4kT_{n0} = 2Gw_v+2(w_iw_v+G^2w_v^2)^{\frac{1}{2}} \tag{12.38}$$

and

$$R_{s0} = \left(\frac{w_v}{w_i+G^2w_v}\right)^{\frac{1}{2}}. \tag{12.39}$$

We now give one example of the use of these results and consider a bipolar transistor at low frequencies with a d.c. current gain β_{dc} and a series base resistance R_b. We shall assume that the input conductance of the ideal device g_m/β is negligible so that $G = 0$ and

$$w_v = 2kT/g_m + 4kTR_b,$$

$$w_i = 2kTg_m/\beta_{dc}.$$

This gives

$$R_{s0} = \frac{\beta_{dc}^{\frac{1}{2}}}{g_m}(1+2g_mR_b)^{\frac{1}{2}}$$

and

$$T_n = T\beta_{dc}^{-\frac{1}{2}}(1+2g_mR_b)^{\frac{1}{2}},$$

while the noise figure F_0 referred to a source at the same temperature T as the transistor is

$$F_0 = 1+\beta_{dc}^{-\frac{1}{2}}(1+2g_mR_b)^{\frac{1}{2}}.$$

If, for example, $\beta_{dc} = 100$, $R_b = 250\ \Omega$, and the collector current is 50 μA so that $g_m = 2$ mA V^{-1} we obtain $R_{s0} \sim 7000\ \Omega$ and $F_0 = 1{\cdot}14$. We note that $\beta/g_m = 50\ 000\ \Omega$, so that we were justified in neglecting G since G^{-1} is greater than both R_{s0} and R_b. This noise figure of about $\frac{1}{2}$ dB corresponds to about the optimum performance attainable with currently available transistors.

12.4. Cascaded stages

If, as in the last section, we take the output of a single stage to be the drain, collector, or anode current we have ignored any Johnson noise generated in its load resistance. Fig. 12.2 illustrates a complete amplifier with its load

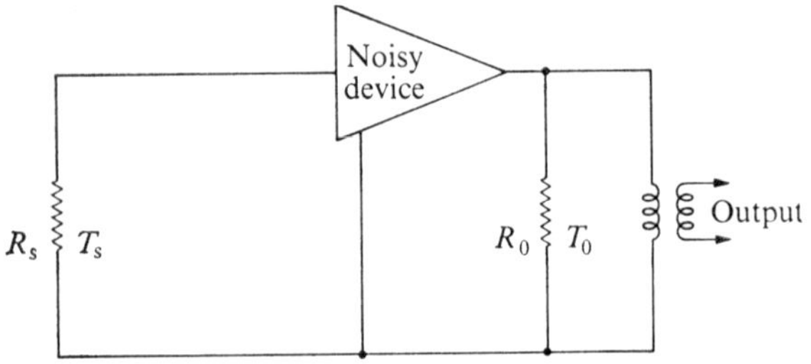

FIG. 12.2 A single amplifier stage with a source at T_s and a load of T_0.

resistance R_0 chosen to optimize its performance, but followed by an ideal transformer to present a suitable source impedance for the next stage. If a_1 is the power gain of this stage the available noise power at the output terminals has a power spectrum

$$w_1 = a_1FkT_s + kT_0, \tag{12.40}$$

where F is the noise figure of the device itself. The over-all noise figure is

$$F_1 = F + T_0/a_1 T_s. \tag{12.41}$$

and is approximately equal to F only if the gain a_1 is large. The over-all noise temperature of this system is

$$T_1 = (F_1 - 1)T_s = (F-1)T_s + T_0/a_1. \tag{12.42}$$

If the output of this stage is connected to a second stage whose noise temperature, including the effect of its load, is T_2 and whose gain is a_2, the output noise power spectrum is

$$w_2 = a_2(w_1 + kT_2) = a_1 a_2(kT_s + kT_1) + a_2 kT_2, \tag{12.43}$$

and the over-all noise temperature is

$$T_{12} = T_1 + T_2/a_1. \tag{12.44}$$

Clearly, if a_1 is large enough the noise in the second stage is negligible.

If the stages are reversed in order, the over-all noise temperature is

$$T_{21} = T_2 + T_1/a_2, \tag{12.45}$$

and this will be less than T_{12} if

$$T_2 + \frac{T_1}{a_2} < T_1 + \frac{T_2}{a_1}, \tag{12.45}$$

or

$$\frac{T_2}{1 - 1/a_2} < \frac{T_1}{1 - 1/a_1}. \tag{12.46}$$

Thus the choice of which amplifier to put first depends not only on its noise temperature but also to some extent on its gain, and to express this relationship Haus and Adler (1959) introduced the concept of a noise measure

$$M = \frac{F-1}{1-1/a} \tag{12.47}$$

as an over-all figure of merit. We might use instead the measure

$$\theta = \frac{T_n}{1-1/a}. \tag{12.48}$$

If N stages with noise temperature T_1, T_2 ... and gains a_1, a_2, etc. are cascaded the output noise power spectrum is

$$W = k\{(T_s + T_1)a_1 a_2 \ldots a_N + T_2 a_2 \ldots a_N + \ldots + T_N a_N\},$$

and the over-all noise temperature is

$$T_n = T_1 + T_2/a_1 + \ldots + T_N/a_1 a_2 \ldots a_{N-1}. \qquad (12.49)$$

For a cascade of N identical stages

$$T_n = T_1\{1+1/a+\ldots+(1/a)^{N-1}\} = T_1 \frac{1-(1/a)^N}{1-1/a} \sim \frac{T_1}{1-1/a} = \theta, \qquad (12.50)$$

and similarly

$$F-1 = \frac{F_1-1}{1-1/a} = M. \qquad (12.51)$$

Thus the over-all noise temperature or noise figure of a high-gain amplifier is equal to the noise measures of the individual stages.

When stages with different noise measures are used in a cascade, the over-all noise temperature is least when the stages are used in order of increasing noise measure, and the over-all noise temperature, obtained from (12.49) by putting $T_1 = \theta_1[1-(1/a_1)]$, etc. is

$$T_n = \theta_1 + \frac{\theta_2-\theta_1}{a_1} + \frac{\theta_3-\theta_2}{a_1 a_2} + \text{etc.} \qquad (12.52)$$

and so T_n can never be less than the noise measure θ of the first stage. There is little point in attempting to reduce the noise temperature of this stage if in the process the gain is so far reduced that the noise measure is increased.

12.5. Negative feedback

Negative feedback is often employed to modify the gain and input admittance of an amplifier, and we must therefore consider its effect on the noise performance. We shall see that it is generally small except when the gain of the system within the feedback loop is low.

Fig. 12.3 shows a source resistance R_s connected to an amplifier of voltage gain $-A$ and input conductance G, whose properties have been modified by

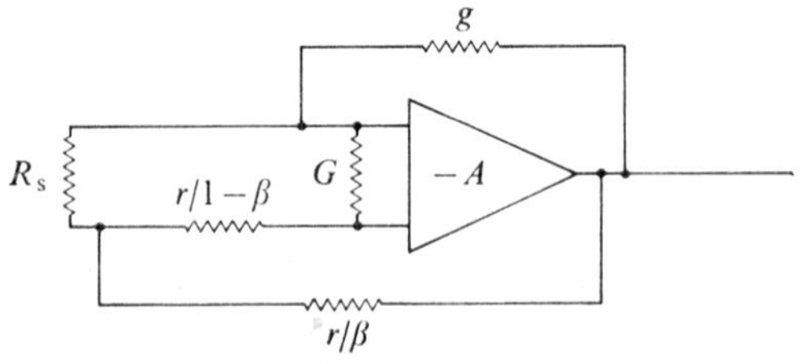

FIG. 12.3. Amplifier with negative voltage and current feedback.

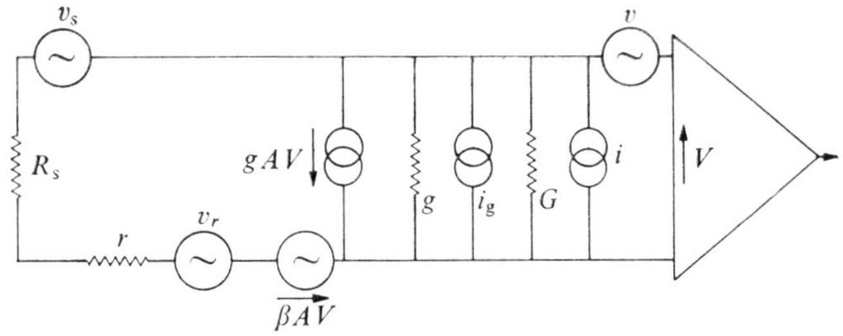

FIG. 12.4. Equivalent circuit including noise generators for an amplifier with negative feedback.

the addition of a voltage, or series, feedback loop and a current, or shunt, feedback loop.

In Fig. 12.4 the feedback elements are represented in the equivalent circuit by r and its associated voltage generator $-\beta AV$ and by g and its current generator $-gAV$. The figure also includes the noise generators v and i associated with the amplifier and also the Johnson noise generators v_s, v_r, and i_g associated with R_s, r, and g. It is now straightforward to calculate the voltage V across the input of the ideal amplifier, and it is

$$V = \left\{1+\frac{\beta A+gA(R_s+r)}{1+(G+g)(R_s+r)}\right\}^{-1}\left\{v+\frac{(i+i_g)(R_s+r)+v_s+v_r}{1+(G+g)(R_s+r)}\right\}.$$

The noise figure is determined by the ratio of the terms depending on i, v, i_g, and v_g to the term arising from Johnson noise in R_s. Since all the sources are uncorrelated we have to consider the sum of the squares of the individual terms and, in terms of the power spectra, we obtain

$$F-1 = \frac{(R_s+r)^2 w_i+\{1+(G+g)(R_s+r)\}^2 w_v+(R_s+r)^2 w_g+w_r}{w_s}.$$

We shall assume that r and g are at the same temperature T, so that $w_r = 4kTr$ and $w_g = 4kTg$. At the same time we express w_i as $4kTg_n$ and w_v as $4kTr_n$ and, of course, $w_s = 4kT_sR_s$. We also replace $F-1$ by T_n/T_s and thus obtain

$$\frac{T_n}{T_s} = \frac{(R_s+r)^2 g_n}{R_s}+\frac{\{1+(G+g)(R_s+r)\}^2 r_n}{R_s}+\frac{(R_s+r)^2 g}{R_s}+\frac{r}{R_s}. \quad (12.53)$$

Without feedback

$$\frac{T_n}{T_s} = R_s g_n+\frac{(1+GR_s)^2 r_n}{R_s}, \quad (12.54)$$

and so we see that feedback always increases the noise. This is true even if we omit the last two terms in (12.53), which arise from Johnson noise in the feedback components. Thus even if these components were purely reactive the noise temperature would still be increased. In practice, $(G+g)(R_s+r)$ is almost always small, and we can neglect the corresponding term in (12.53) which then becomes

$$\frac{T_n}{T} = 2g_n r + 2gr + R_s(g_n+g) + R_s^{-1}(r_n + r + r^2 g_n + r^2 g). \qquad (12.55)$$

With voltage feedback alone, i.e. with $g = 0$, this reduces to

$$\frac{T_n}{T} = 2g_n r + R_s g_n + R_s^{-1}(r_n + r + r^2 g_n), \qquad (12.56)$$

and the increased optimum source resistance is

$$R_{s0} = \left(\frac{r_n}{g_n}\right)^{\frac{1}{2}} \left(1 + \frac{r}{r_n} + r^2 \frac{g_n}{r_n}\right)^{\frac{1}{2}} \qquad (12.57)$$

and the optimum noise temperature is

$$T_n = 2T\left\{g_n r + (g_n r_n)^{\frac{1}{2}} \left(1 + \frac{r}{r_n} + r^2 \frac{g_n}{r_n}\right)^{\frac{1}{2}}\right\}. \qquad (12.58)$$

If, for example, we consider an amplifier with $r_n = 100\ \Omega$, $g_n = 10^{-4}\ \Omega^{-1}$, and $r = 1000\ \Omega$ the optimum values without feedback are $R_{s0} = 1000\ \Omega$ and $T_n = T/5$. With feedback $R_{s0} = 3500\ \Omega$ and $T_n \sim T$. This example might correspond perhaps to a bipolar transistor with $I_c = 250\ \mu A$ and a resistance of $1000\ \Omega$ in the emitter lead.

With $r = 0$ and current feedback alone we have

$$\frac{T_n}{T} = R_s(g_n + g) + r_n R_s^{-1}, \qquad (12.59)$$

and the reduced optimum source resistance is

$$R_{s0} = \left(\frac{r_n}{g_n}\right)^{\frac{1}{2}} \left(1 + \frac{g}{g_n}\right)^{-\frac{1}{2}}, \qquad (12.60)$$

while the increased optimum noise temperature is

$$T_{n0} = 2T(r_n g_n)^{\frac{1}{2}} \left(1 + \frac{g}{g_n}\right)^{\frac{1}{2}}. \qquad (12.61)$$

A junction f.e.t. operating at 100 kHz might have $r_n \sim 200\ \Omega$ and

$$g_n \sim 2 \times 10^{-8}\ \Omega^{-1}.$$

Without feedback, $R_{s0} = 10^5\ \Omega$ and $T_n = 4 \times 10^{-3}\ T$. If a resistance $R = 10^4\ \Omega$ (so that $g = 10^{-4}\ \Omega^{-1}$) is connected from the drain to the gate to give negative current feedback, R_{s0} is reduced by a factor $\frac{1}{70}$ and T_n is increased by 70.

If feedback is applied to a virtually noiseless amplifier so that g_n and r_n are negligible in (12.53) we have

$$T_n/T = 2gr + R_s g + R_s^{-1} r(1+gr);$$

for voltage feedback alone

$$T_n/T \sim r/R_s, \tag{12.62a}$$

and for current feedback alone

$$T_n/T \sim gR_s. \tag{12.62b}$$

If these two formulae lead to values of T_n greater than those expected for the amplifier without feedback it is some indication that feedback will have a significant effect on the noise. Thus, for example, if an amplifier without feedback has an optimum source resistance of 500 Ω and leads to $T_n/T = \frac{1}{2}$ the insertion of a feedback resistance $r = 500$ Ω will at least double the noise.

12.6. Different device configurations

So far we have assumed implicitly that the active device is used in the normal grounded-emitter, source, or cathode configuration. We now consider other configurations, e.g. the grounded-base configuration or the emitter-follower.

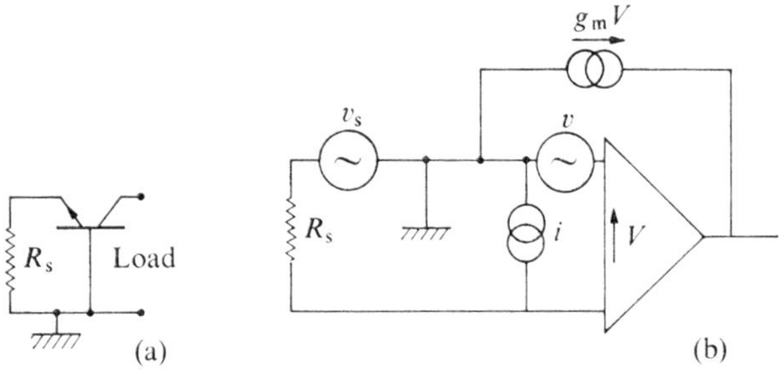

FIG. 12.5(a) and (b). Grounded base stage and equivalent circuit.

For simplicity we continue to assume that the device in the normal configuration has a negligible input admittance. Fig. 12.5(a) shows a grounded-base stage and Fig. 12.5(b) shows the noise equivalent circuit. We have

$$V = v + iR_s + v_s - g_m R_s V,$$

and so the power spectrum of V is

$$W = \left(\frac{1}{1+g_m R_s}\right)^2 (4kTR_s + 4kTr_n + 4kTR_s^2 g_n).$$

This gives a noise temperature

$$T_n = T\left(\frac{r_n}{R_s} + R_s g_n\right),$$

and both $R_{s0} = (r_n/g_n)^{\frac{1}{2}}$ and the minimum noise temperature $T_{n0} = 2T(r_n g_n)^{\frac{1}{2}}$ are the same as in the grounded-emitter configuration. The input admittance for this configuration is, however, high and equal to g_m. Thus, whereas for the normal configuration R_{s0} is usually much less than the input impedance G^{-1}, in the grounded-base configuration R_{s0} is greater than the input impedance, which is now g_m^{-1}. In neither case is R_{s0} equal to the source impedance giving maximum gain.

Fig. 12.6 shows the equivalent circuit for the emitter, source, or cathode

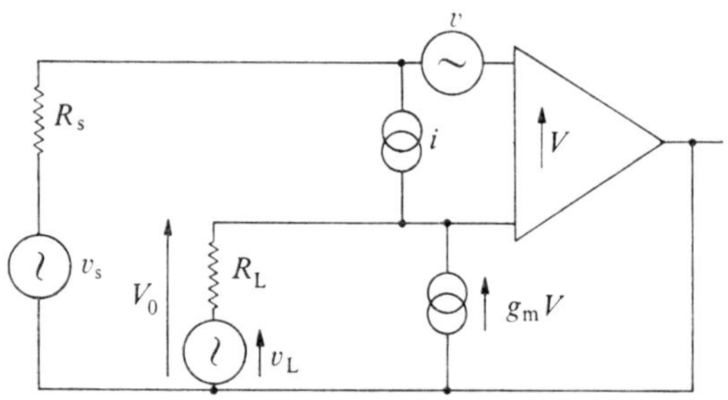

FIG. 12.6. Equivalent circuit of a source or emitter follower with a load R_L. The output voltage is V_0.

follower, and we must now include noise generated in the load R_L. We have

$$V(1+g_m R_L) = v+v_s+iR_s-v_L$$

and

$$V_0 = v_L+g_m R_L V,$$

so that

$$V_0 = \frac{v_L+g_m R_L(v+v_s+iR_s)}{1+g_m R_L}.$$

The noise temperature is then given by

$$T_n = T\left\{\frac{r_n}{R_s}+g_n R_s+\frac{R_L}{R_s}(g_m R_L)^{-2}\right\}. \tag{12.63}$$

If $g_m R_L$ is large this is the same result as for the other configurations.

The input impedance is now much higher than R_{s0} and the output impedance is low and equal to $1/g_m$. If however the circuit is matched to a load $R_L = 1/g_m$ the noise temperature becomes

$$T_n = T\left(\frac{r_n}{R_s}+g_n R_s+\frac{1}{g_m R_s}\right)$$

and the minimum value is

$$T_{n0} = 2T(g_n r_n)^{\frac{1}{2}}\left(1+\frac{1}{g_m r_n}\right)^{\frac{1}{2}}. \tag{12.64}$$

If the active device is a bipolar transistor run at such a low collector current that we can neglect the effect of the series base resistance the noise equivalent resistance is $r_n = \frac{1}{2}g_m$ and so, with a matched load, the noise is increased by a factor $\sqrt{3}$.

12.7. Impedance matched input stages

The source impedance R_{s0} that optimizes the noise performance of an input stage is not usually equal to the input impedance of the stage. In some applications, however, it is essential that the amplifier present the correct terminating impedance to the source, e.g. in pulse amplifiers connected to a source by a matched transmission line, and it is worth seeing how this can be achieved and what it costs in noise performance.

At not too high frequencies it is possible to apply negative feedback around several stages with a high over-all gain; this need not lead to any deterioration in the noise performance but can be used to make the input impedance equal to R_{s0}. This is therefore one possible approach. At higher frequencies, however, feedback round several stages is likely to lead to instability, and this technique is not available.

If R_{s0} is the optimum source impedance and T_{n0} the optimized noise temperature, the value of T_n for other source impedances is given by

$$T_n - 2Gr_nT = \tfrac{1}{2}(T_{n0} - 2Gr_nT)\left(\frac{R_{s0}}{R_s}+\frac{R_s}{R_{s0}}\right), \tag{12.65}$$

where G is the input admittance of the device and r_n is its equivalent noise resistance at a temperature T. If we match the source to the device so that $R_sG = 1$, we have

$$T_n - 2Gr_nT = \tfrac{1}{2}(T_{n0} - 2Gr_nT)(GR_{s0}+1/GR_{s0}). \tag{12.66}$$

Consider, for example, an f.e.t. operating at 50 MHz, for which we might have $r_n = 200\ \Omega$ with $T = 300$ K and $G = 3\times10^{-4}\ \Omega^{-1}$, while $R_{s0} = 1600\ \Omega$, so that $GR_{s0} \sim \frac{1}{2}$. The quantity $\frac{1}{2}(GR_{s0}+1/GR_{s0})$ is then only 1·25, and the change in T_n from its optimum value is negligible. On the other hand, for a bipolar transistor, where GR_{s0} is of the order $\beta^{-\frac{1}{2}}$, the value of

$$\tfrac{1}{2}(GR_{s0}+1/GR_{s0}) \sim \tfrac{1}{2}\beta^{\frac{1}{2}} \sim 5.$$

Clearly, it may sometimes be possible to match the source to the input without a disastrous increase in T_n, but in other cases it may not be possible.

When the input impedance G^{-1} is greater than R_{s0} it is common practice to place a resistance $R \sim R_s$ across the source and then transform the resulting resistance $R_{s/2}$ to equal R_{s0}. Conversely, when G^{-1} is much less than R_{s0} a resistance is placed in series with the source and the resulting resistance $2R_s$ transformed to equal R_{s0}. Both cases lead to identical results and we shall only consider the first. The equivalent circuit is shown in Fig. 12.7. The

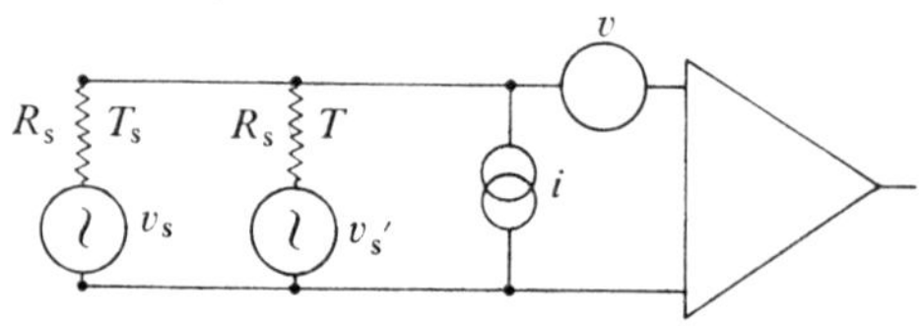

FIG. 12.7. Amplifier input stage with a matching resistor R_s at T.

amplifier noise is minimized when $R_s/2$ has been transformed to equal $R_{s0} = (w_v/w_i)^{\frac{1}{2}}$, and then the minimum noise temperature is

$$T_n' = T + 2T_{n0}, \tag{12.67}$$

where T_{n0} refers to the optimized amplifier. Thus, for example, if $T = 300$ K and $T_{n0} = 150$ K the noise temperature is raised, by this expedient, from 150 K to 600 K.

In some cases, it may be difficult to transform the source impedance so that the parallel combination of resistance $R_s/2$ is equal to R_{s0}. In this case we have

$$T_n' = T + T_{n0}\left(\frac{R_s}{2R_{s0}} + \frac{2R_{s0}}{R_s}\right). \tag{12.68}$$

The increase in noise may then be very considerable. Suppose, for example, that the source is a 50 Ω cable so that $R_s = 50$ Ω and that the input stage is a bipolar transistor requiring $R_{s0} = 500$ Ω to give a minimum $T_{n0} = 300$ K, then we have

$$T_n = (300 + 3000)\text{ K} = 6300\text{ K}.$$

Obviously this simple technique is unsatisfactory.

At low frequencies, without encountering instability, we can alter the input impedance by using negative feedback around several stages of high gain, and this will cause no appreciable deterioration in the noise performance. Usually at high frequencies we are restricted to enclosing only one stage within the feedback loop. Consider then a high-impedance stage of voltage gain $-A$, whose input impedance is reduced to $R = 1/Ag$ by the addition of a current-feedback loop. According to eqn (12.59),

$$T_n'/T = R_s(g_n + g) + r_n R_s^{-1},$$

and if we make $R = R_s$ by choosing g to equal $1/AR_s$ we have

$$T'_n/T = R_s g_n + r_n R_s^{-1} + 1/A.$$

If R_s can be transformed to equal $R_{s0} = (r_n/g_n)^{\frac{1}{2}}$ we have, instead of $T_n = T_{n0} = 2(g_n r_n)^{\frac{1}{2}}T$, $T_n = T_{n0} + T/A$. If, for example, $T_{n0} = 150$ K, $T = 300$ K, and $A = 10$, the increase in T_n is negligible. Even if $A = 2$ it is less than any scheme without feedback. If R_s cannot be transformed we have

$$T'_n = \frac{1}{2}\left(\frac{R_{s0}}{R_s} + \frac{R_s}{R_{s0}}\right)T_{n0} + \frac{T}{A}, \tag{12.69}$$

which we may compare with (12.68). With the same numerical values $R_s = 50\ \Omega$, $R_{s0} = 500\ \Omega$, $T_{n0} = 300$ K, and $T = 300$ K, and with $A = 2$ it yields $T_n = 1650$ K which, though not low, is better than 6300 K.

It will be apparent that, especially at high frequencies, it is rather difficult to achieve good noise performance and at the same time match an amplifier to a source.

12.8. Differential sensitivity

The signal input power that yields unity output signal to noise ratio with an amplifier of noise figure F and bandwidth Δf is $FkT\Delta f$, but in some cases we are interested in the smallest detectable change δP in a relatively large power P. This we will assume to be the change that results in an amplitude change equal to the r.m.s. noise $(FkT\Delta f)^{\frac{1}{2}}$, and so we require

$$\delta P \frac{\mathrm{d}(P)^{\frac{1}{2}}}{\mathrm{d}P} = (FkT\Delta f)^{\frac{1}{2}},$$

which yields

$$\delta P = 2(PFkT\Delta f)^{\frac{1}{2}}. \tag{12.70}$$

Consider, for example, an a.c. bridge driven by a voltage of r.m.s. value V_0, which produces an out of balance voltage ηV_0, where η depends on the bridge setting. If the noise figure and bandwidth of the amplifier connected to the output are F and Δf and the output impedance of the bridge is R, then the minimum detectable change in η is given by

$$V_0\,\delta\eta = (4RFkT\Delta f)^{\frac{1}{2}}.$$

But we also have

$$\delta P = \frac{1}{4R}\,\delta(\eta V_0)^2 = \frac{\eta V_0^2\,\delta\eta}{2R} = 2P\frac{\delta\eta}{\eta},$$

and so either

$$\delta\eta = \left(\frac{4RFkT\Delta f}{V_0^2}\right)^{\frac{1}{2}}$$

or

$$\delta\eta = \left(\frac{\eta^2 FkT\Delta f}{P}\right)^{\frac{1}{2}}.$$

These two results are clearly the same. If F and Δf do not vary with the out of balance output power P, the sensitivity to small changes in the balance parameter η is independent of the value of η.

If, on the other hand, F or Δf changes with the input applied to the amplifier the sensitivity can be optimized by choosing a particular off-balance output. This, as we shall see, is of some importance in paramagnetic resonance spectrometers (see Chapter 22).

12.9. Synopsis

The mean square output fluctuations of an amplifier can be expressed in terms of the power spectrum $W(f)$ of the output as

$$\langle\Delta\theta^2\rangle = \int_0^\infty W(f)\,\mathrm{d}f.$$

The output noise $W(f)$ of an amplifier of gain $A(f)$ connected to a white noise source with a power spectrum w_s contains a contribution $A(f)A^*(f)w_s$ due to the source and an additional contribution arising from internal sources. The spot-frequency noise figure relative to this source is $F(f)$, defined so that

$$W(f) = F(f)A(f)A^*(f)w_s.$$

The effective bandwidth Δf is defined as

$$\Delta f = \frac{\int_0^\infty A(f)A^*(f)\,\mathrm{d}f}{A_0A_0^*},$$

where A_0 is the gain within the pass-band, and the average noise figure $\bar{F}$ is defined so that

$$\langle\Delta\theta^2\rangle = \int_0^\infty W(f)\,\mathrm{d}f = \bar{F}A_0A_0^*w_s\,\Delta f.$$

This is convenient because, in many cases, Δf is determined by the later stages of the amplifier while $F(f)$ and $\bar{F}$ are determined by the earlier stages, and the two parameters F and Δf can be discussed separately.

If w_s is related to the source temperature T_s by $w_s = kT_s$ we can express F as

$$F = 1+T_n/T_s,$$

where T_n is the noise temperature of the amplifier. Usually F is quoted relative to a source at 293 K and so

$$T_n = 293(F-1).$$

If an amplifier is constructed using stages each of gain G' and noise figure F' the over-all noise figure is

$$F = 1+\frac{F'-1}{1-1/G'},$$

and the quantity

$$M' = \frac{F'-1}{1-1/G'}$$

is known as the noise measure of the stages. An amplifier constructed of stages of noise measures M_1, M_2 etc. cannot have a noise figure less than the least of these measures, and will approach this limit most closely if the stage of lowest measure is placed at the input.

For any given input stage there is a source impedance which minimizes the noise figure. Usually this, in conjunction with the input susceptance of the stage, will be real or approximately so. If we denote it by R_{s0} and the optimized noise figure by F_0, then, for other values of source resistance,

$$F = 1+a+\tfrac{1}{2}(F_0-1-a)\left(\frac{R_s}{R_{s0}}+\frac{R_{s0}}{R_s}\right),$$

where a is a constant, usually small, and determined mainly by the ratio of R_{s0} to the real input impedance of the stage. When a is zero and the noise generators of the input stage are described in terms of a noise conductance g_n and noise resistance r_n we have $R_{s0}^2 = r_n/g_n$ and $F_0 = 1+2(r_n g_n)^{\frac{1}{2}}$. The effects of a finite input impedance tend to increase F_0.

Correlations between the two amplifier noise generators have a negligible effect on R_{s0} and, if the input circuit is made slightly reactive (usually capacitative), can lead to a small decrease in T_{n0}. Negative feedback tends to increase T_n, but if the gain within the feedback loop is large substantial changes in the gain and input admittance can be achieved without an appreciable increase in T_n. Different configurations of the same device yield the same value of T_{n0} and require the same value of R_{s0}.

In general the lowest noise temperature is not achieved when the device is matched to the source and the gain is maximized. It is difficult, even using negative feedback, to make these two conditions coincide without substantially increasing T_n.

13 Transistor Amplifiers

AT frequencies above about 1 GHz the most satisfactory low-noise amplifiers are the travelling-wave tube, the parametric amplifier, and the maser (to be dealt with in later chapters); below about 1 kHz the noise performance of amplifiers is dominated by flicker noise. This low-frequency region will be considered in the next chapter, but here we confine ourselves to moderate frequences in the range from about 1 kHz to 1 GHz.

In f.e.t.s the equivalent voltage noise at the gate terminal does not change appreciably with frequency but the gate-current noise consists of two components; one has a power spectrum $2eI_g$ which is independent of frequency and is due to gate leakage current, the other has a power spectrum $g_m kT(f/f_T)^2$, is due to induced gate noise, and increases quadratically with frequency. The two terms are equal when

$$\frac{f}{f_T} \sim \left(\frac{2eI_g}{g_m kT}\right)^{\frac{1}{2}};$$

with $I_g = 10^{-9}$ A, $g_m = 5$ mA V^{-1} and $f_T = 100$ MHz this occurs at about $\frac{1}{2}$ MHz.

In bipolar transistors the equivalent noise voltage at the base terminal is similarly independent of frequency up to about $f_T/2$, but the base-current noise contains a frequency-independent term $2eI_b$ or $2kTg_m/\beta$ and a frequency-dependent term $4kTg_m(f/f_T)^2$. These are equal when $f = f_T(2\beta)^{-\frac{1}{2}}$. With $\beta \sim 200$ this is $f_T/20$ and, depending on the value of f_T, may be anywhere from 100 kHz to 100 MHz.

These two characteristics frequencies serve to define a boundary between a low-frequency regime and a high-frequency regime; we note however that for some bipolar transistors 50 MHz is a low frequency.

13.1. Field-effect transistors

The equivalent input circuit for an f.e.t. is shown in Fig. 13.1. For all practical purposes the two noise generators may be regarded as uncorrelated. The power spectrum of the voltage generator is

$$w_v = \tfrac{2}{3}.4kT/g_m, \tag{13.1}$$

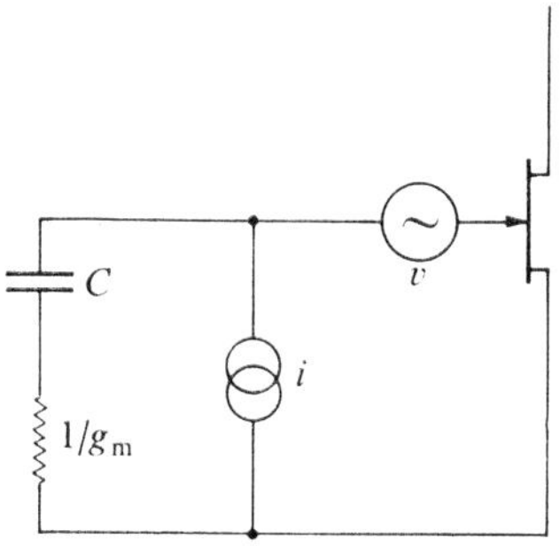

FIG. 13.1 Equivalent input circuit of an f.e.t.

and the equivalent noise resistance is therefore

$$r_n = \tfrac{2}{3}g_m. \tag{13.2}$$

At low frequencies the dominant term in the current noise has a power spectrum

$$w_i = 2eI_g, \tag{13.3}$$

where I_g is the gate leakage current. In this frequency region we can neglect C and the noise temperature with a source resistance R_s is simply

$$T_n = \tfrac{1}{2}T_{n0}\left(\frac{R_s}{R_{s0}}+\frac{R_{s0}}{R_s}\right), \tag{13.4}$$

where

$$R_{s0} = \left(\frac{w_v}{w_i}\right)^{\frac{1}{2}} = \left(\frac{4kT}{3eI_g g_m}\right)^{\frac{1}{2}} \sim \left(\frac{1}{30I_g g_m}\right)^{\frac{1}{2}} \tag{13.5}$$

and

$$T_{n0} = T\left(\frac{4eI_g}{3kTg_m}\right)^{\frac{1}{2}} \sim T\left(\frac{50I_g}{g_m}\right)^{\frac{1}{2}}. \tag{13.6}$$

Thus if $I_g = 10^{-10}$ A and $g_m = 5$ mA V^{-1} we obtain $R_{s0} \sim 2{\cdot}5\times10^5$ and $T_n \sim 10^{-3}\,T \sim 0{\cdot}3$ K. Provided that R_s can be made as large as this, very low noise temperatures are attainable.

If R_s cannot be made large enough we have

$$T_n \sim \tfrac{1}{2}T_{n0}\frac{R_{s0}}{R_s} = T\frac{r_n}{R_s} = \frac{2T}{3g_m R_s}. \tag{13.7}$$

For example, if $g_m = 5$ mA V^{-1} and $R_s = 10^4\ \Omega$ then, with $T = 300$ K, we obtain $T_n = 4$ K, which for most purposes is negligible.

At higher frequencies we have to add the induced gate noise term to w_i. This is

$$w_i = \frac{\omega^2 c^2}{4g_m}.4kT = \frac{1}{4}\left(\frac{f}{f_T}\right)^2 4kTg_m, \tag{13.8}$$

where

$$f_T = g_m/2\pi C. \tag{13.9}$$

is the gain–bandwidth product. If we express w_i as $4kTg_n$ we have

$$g_n = \tfrac{1}{4}g_m\left(\frac{f}{f_T}\right)^2. \tag{13.10}$$

At frequencies where this term is dominant, but we can still ignore the input conductance which is

$$G = \frac{\omega^2 c^2/g_m}{1+\omega^2 c^2/g_m^2} = \frac{(f/f_T)^2}{1+(f/f_T)^2}g_m, \tag{13.11}$$

we have

$$R_{s0} = \left(\frac{r_n}{g_n}\right)^{\frac{1}{2}} = \left(\frac{8}{3}\right)^{\frac{1}{2}}\frac{f_T}{fg_m}, \tag{13.12}$$

and

$$T_{n0} = 2T(r_n g_n)^{\frac{1}{2}} = \left(\frac{2}{3}\right)^{\frac{1}{2}}\frac{f}{f_T}T. \tag{13.13}$$

With $f_T = 200$ MHz and $I_g = 10^{-10}$A these formulae would be valid from about 200 kHz to 20 MHz and at, say, 2 MHz with $g_m = 5$ mA V^{-1} would yield $R_{s0} \sim 30$ kΩ and $T_{n0} \sim 3$ K.

At higher frequencies we cannot ignore G, and we have, from eqns (12.38) and (12.39),

$$T_{n0} = 2Gr_nT + 2(g_n r_n + G^2 r_n^2)^{\frac{1}{2}}T$$

and

$$R_{s0} = \left(\frac{r_n}{g_n + G^2 r_n}\right)^{\frac{1}{2}}.$$

These results can be expressed as

$$T_{n0} = \left(\frac{2}{3}\right)^{\frac{1}{2}} T\frac{f}{f_T}\frac{\{1+(14/3)(f/f_T)^2+(f/f_T)^4\}^{\frac{1}{2}}+(8/3)^{\frac{1}{2}}(f/f_T)}{1+(f/f_T)^2} \tag{13.14}$$

and

$$R_{s0} = \frac{1}{g_m}\left(\frac{8}{3}\right)^{\frac{1}{2}}\frac{f_T}{f}\frac{1+(f/f_T)^2}{1+(14/3)(f/f_T)^2+(f/f_T)^4\}^{\frac{1}{2}}}. \tag{13.15}$$

If, for example, $f = \frac{1}{2}f_T$ we obtain $T_{n0}/T \sim 0{\cdot}75$ and $F = 1{\cdot}75$ or 2·5 dB. This agrees well with experimental values, thus the type BFW 10 with $f_T \sim 200$ MHz is quoted by the manufacturers as having a noise figure of 2·5 dB at 100 MHz.

The noise equivalent circuit of a vacuum tube is shown in Fig. 13.2.

If we replace w_v by

$$w_v = 2.5\frac{4kT}{g_m},$$

so that

$$r_n = 2.5/g_m,$$

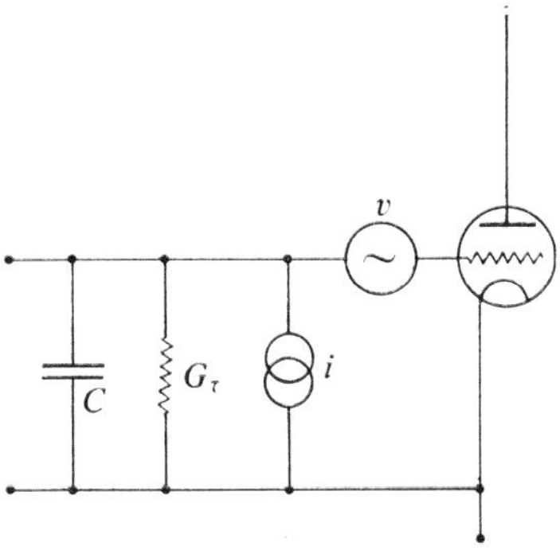

FIG. 13.2. Triode input equivalent circuit.

and w_i by

$$w_{\mathrm{i}} = 2eI_{\mathrm{g}} + 4kT(5G_{\tau}),$$

where G_τ is the transit-time conductance proportional to f^2, the results are similar to those for f.e.t.s (but worse). A noise temperature $T_{\mathrm{n0}} = 300$ K and $F = 2$ or 3 dB at 100 MHz would be regarded as exceptionally good.

13.2. Bipolar transistors

At frequencies well below f_T the noise equivalent circuit of a bipolar transistor connected to a source R_{s} at T_{s} has the form shown in Fig. 13.3. The series base resistance is R_{b} and the power spectra of the four noise sources are

$$w_v = 2kT/g_{\mathrm{m}} = 4kTr_{\mathrm{n}}, \tag{13.16a}$$

$$w_i = 2kTg_{\mathrm{m}}/\beta_{\mathrm{dc}} = 4kTg_{\mathrm{n}}, \tag{13.16b}$$

$$w_{\mathrm{b}} = 4kTR_{\mathrm{b}}, \tag{13.16c}$$

$$w_{\mathrm{s}} = 4kT_{\mathrm{s}}R_{\mathrm{s}}, \tag{13.16d}$$

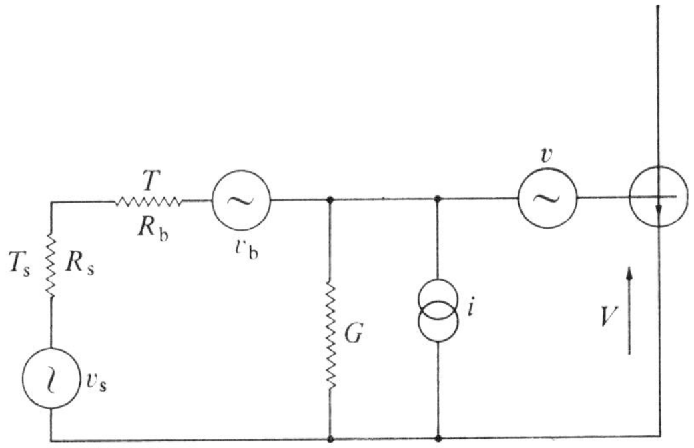

FIG. 13.3. Equivalent circuit for a bipolar transistor connected to a source R_{s} at T .

while the input conductance is

$$G = g_m/\beta = eI_c/\beta kT. \qquad (13.17)$$

If we ignore R_b and G the noise figure is

$$F = 1+\left(\frac{r_n}{R_s}+g_n R_s\right)\frac{T}{T_s} \qquad (13.18)$$

and

$$T_n = T\left(\frac{r_n}{R_s}+g_n R_s\right), \qquad (13.19)$$

which has a minimum value

$$T_{n0} = 2(r_n g_n)^{\frac{1}{2}} = \beta_{dc}^{-\frac{1}{2}}T \qquad (13.20)$$

when

$$R_{s0} = \left(\frac{r_n}{g_n}\right)^{\frac{1}{2}} = g_m^{-1}\beta_{dc}^{\frac{1}{2}}. \qquad (13.21)$$

If, for example, $I_c = 100\ \mu A$ so that $g_m^{-1} = 250\ \Omega$ and $\beta_{dc} = 400$, we have $R_{s0} = 5000\ \Omega$ and $T_{n0} = T/20 = 15$ K. We note that R_{s0} is much less than $G^{-1} = \beta g_m^{-1} \sim 100\,000\ \Omega$.

Usually R_b is between 10 Ω and 1000 Ω, and we can neglect the effects of G, but we cannot ignore R_b in comparison with $r_n = \frac{1}{2}g_m$. Thus if we retain R_b but ignore G we obtain

$$(F-1)T_s = T_n = T\left\{\frac{R_b+r_n}{R_s}+g_n\frac{(R_s+R_b)^2}{R_s}\right\}. \qquad (13.22)$$

This gives a minimum value

$$T_{n0} = T\left\{\beta_{dc}^{-\frac{1}{2}}\left(1+2R_b g_m+R_b^2\frac{g_m^2}{\beta_{dc}}\right)^{\frac{1}{2}}+R_b\frac{g_m}{\beta_{dc}}\right\} \qquad (13.23)$$

when

$$R_{s0} = \frac{\beta_{dc}^{\frac{1}{2}}}{g_m}\left(1+2R_b g_m+R_b^2\frac{g_m^2}{\beta_{dc}}\right)^{\frac{1}{2}}. \qquad (13.24)$$

The terms involving $R_b g_m/\beta_{dc}$ are almost always negligible, and then

$$T_{n0} = T\beta_{dc}^{-\frac{1}{2}}(1+2R_b g_m)^{\frac{1}{2}}, \qquad (13.25)$$

with

$$R_{s0} = \frac{\beta_{dc}^{\frac{1}{2}}}{g_m}(1+2R_b g_m)^{\frac{1}{2}}. \qquad (13.26)$$

If, for example, $R_b = 200\ \Omega$ and $I_c = 1$ mA so that $g_m = 40$ mA V^{-1}, we have $2R_b g_m = 16$ and both T_{n0} and R_{s0} are increased by a factor 4 by the presence R_b. Thus with $\beta_{dc} = 200$ we get $T_{n0} = 80$ K and $F = 1{\cdot}3$ (or 1 dB).

If I_c is reduced to 100 μA, T_{n0} is reduced to 35 K and F to 1·12 or about $\frac{1}{2}$ dB. This is about the best noise figure obtained in practice. Further reduction of I_c emphasizes the importance of recombination currents in the emitter–base junction and decreases β_{dc}.

The bipolar transistor exhibits a unique and exceedingly convenient feature in that we can alter g_m by changing I_c and so alter the required value of R_{s0}. If we express (13.22) as

$$T_n = T\left\{\frac{R_b}{R_s}+\frac{1}{2g_mR_s}+\frac{g_m}{2\beta_{dc}}\frac{(R_s+R_b)^2}{R_s}\right\}, \tag{13.27}$$

regard R_b and R_s as fixed, and optimize T_n by changing g_m, we find that the optimum value of g_m is

$$g_{m0} = \frac{\beta_{dc}^{\frac{1}{2}}}{R_s+R_b} \tag{13.28}$$

and then

$$T_n = T\left\{\frac{R_b}{R_s}+\beta_{dc}^{-\frac{1}{2}}\frac{(R_b+R_s)}{R_s}\right\} \sim T\left\{\beta_{dc}^{-\frac{1}{2}}+\frac{R_b}{R_s}\right\}. \tag{13.29}$$

For example, if $R_b = 200\ \Omega$ and $R_s = 2000\ \Omega$, while $\beta_{dc} = 100$, we can obtain $T_{n0} = T/5$ simply by adjusting the transistor bias current so that $g_m = 1/220$, which requires $I_c \sim 100\ \mu$A.

At higher frequencies we have to replace the two power spectra w_v and w_i by

$$w_v = \frac{2Tk}{g_m}\{1+(f/f_T)^2\} \tag{13.30a}$$

and

$$w_i = 2kTg_m\,\{\beta_{dc}^{-1}+2(f/f_T)^2\}. \tag{13.30b}$$

High-frequency effects first become noticeable when

$$f \sim f_T(2\beta_{dc})^{\frac{1}{2}} \sim f_T/20. \tag{13.31}$$

When f is appreciably greater than this but still much less than f_T we can obtain the noise simply by replacing β_{dc}^{-1} by $2(f/f_T)^2$ in eqns (13.23)–(13.26). Thus

$$R_{s0} \sim \frac{f_T}{fg_m}(\tfrac{1}{2}+R_bg_m)^{\frac{1}{2}} \tag{13.32}$$

and

$$T_{n0} \sim 2T\frac{f}{f_T}(\tfrac{1}{2}+R_bg_m)^{\frac{1}{2}}. \tag{13.33}$$

If, for example, $R_b = 100\ \Omega$, $I_c = 1$ mA, $g_m = 40$ mA V^{-1}, $f_T = 1000$ MHz, and $f = 100$ MHz, we obtain $T_{n0}/T \sim \frac{1}{2}$ and $F = 1{\cdot}5$ or about 1·5 dB.

At high frequencies the input admittance is mainly capacitative and is

$$Y = \mathrm{j}B = \mathrm{j}\omega C = \mathrm{j}\frac{f}{f_T}g_{\mathrm{m}}. \tag{13.34}$$

The equivalent circuit of a tuned input stage is shown in Fig. 13.4. If L is chosen so that $\omega^2 LC = 1$, using the expressions (13.30a) and (13.30b) for

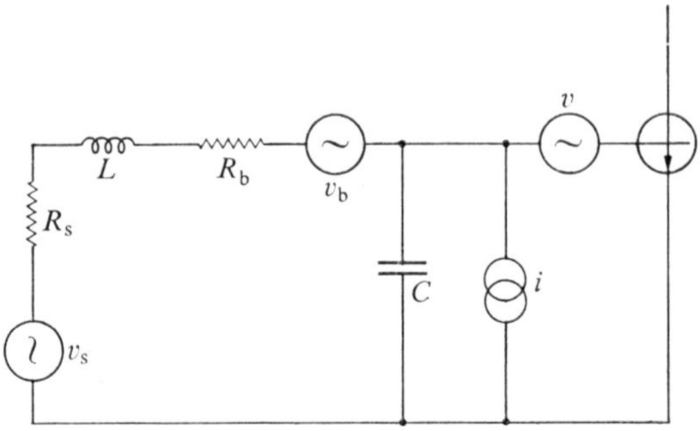

FIG. 13.4. A tuned transistor input stage.

w_v and w_i and setting $\theta = f/f_T$ we obtain

$$\frac{T_{\mathrm{n}}}{T} = g_{\mathrm{m}}R_{\mathrm{b}}\theta^2(1+\theta^2)+\tfrac{1}{2}R_{\mathrm{s}}g_{\mathrm{m}}\theta^2(1+\theta^2)+\frac{1+g_{\mathrm{m}}R_{\mathrm{b}}(1+\theta^2)+\frac{1}{2}g_{\mathrm{m}}^2R_{\mathrm{b}}^2\theta^2(1+\theta^2)}{g_{\mathrm{m}}R_{\mathrm{s}}(1+\theta^2)}.$$

This yields a minimum value

$$\frac{T_{\mathrm{n}}}{T} = \theta\sqrt{2}\{1+g_{\mathrm{m}}R_{\mathrm{b}}(1+\theta^2)+\tfrac{1}{2}g_{\mathrm{m}}^2R_{\mathrm{b}}^2\theta^2(1+\theta^2)\}^{\frac{1}{2}}+g_{\mathrm{m}}R_{\mathrm{b}}\theta^2(1+\theta^2). \tag{13.35}$$

If θ is less than about $\frac{1}{2}$ this is not appreciably different from the result (13.33) obtained by ignoring C. For $\theta = 1$ we have $T_{\mathrm{no}}/T = \sqrt{2}+(2+\sqrt{2})g_{\mathrm{m}}R_{\mathrm{b}}$ and, for example, even if $g_{\mathrm{m}}R_{\mathrm{b}}$ is only unity the noise figure is 5·8 or about 8 dB. Clearly, the base resistance R_{b} exerts an important influence on the noise performance at all frequencies. (See also Appendix B.)

13.3. The cascode circuit

Both types of transistor are rendered potentially unstable by the Miller effect owing to the reverse drain-to-gate or collector-to-base capacitance and, at high frequencies, it is usually necessary to minimize this effect by using a low input-impedance second stage. The cascode circuit, shown in Fig. 13.5(a) employs a second transistor T_2 as a grounded base stage. The first transistor then acts as a high-impedance current source for this stage. The noise equivalent circuit for the second transistor is shown in Fig. 13.5(b). This circuit uses the two current generators i_{b} and i_{c} to represent the device

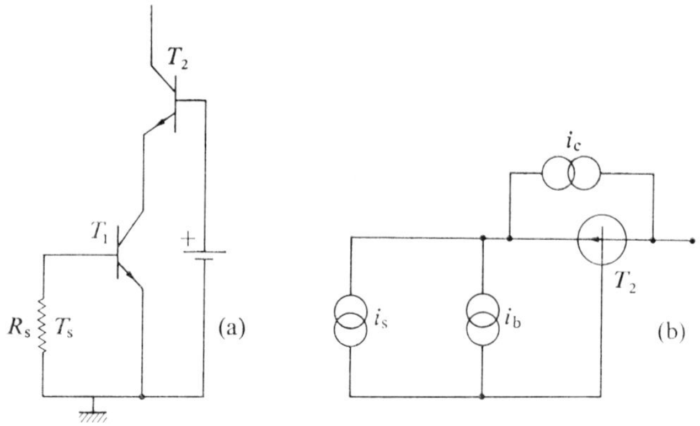

FIG. 13.5(a) and (b). Cascode circuit and the noise equivalent circuit for the second transistor T_2.

noise rather than a voltage generator in the base lead. The output current is simply $i_s+i_b+i_c$ and its power spectrum is $w_s+w_b+w_c$, where

$$w_s = 4F_1kT_sR_sg_m^2,$$

$w_b = 2eI_b$, and $w_c = 2eI_c$. Thus

$$W = 4F_1kT_sR_sg_m^2+2e(I_b+I_c)$$
$$= 4F_1kT_sR_sg_m^2+2eI_e$$

and the over-all noise figure is

$$F = F_1+\frac{2eI_e}{4kT_sR_sg_m^2}. \tag{13.36}$$

Since I_e for the second transistor is equal to I_c for the first transistor and $eI_c = kTg_m$ we have

$$F = F_1+\frac{1}{2R_sg_m}.$$

Further, if R_s is optimized already it is considerably greater than $1/g_m$ and so $F \sim F_1$ and the noise figure of the combination is not appreciably worse than that of T_1 alone.

An alternative circuit which is often more convenient is shown in Fig. 13.6. In this case for the sake of variety we have shown a field-effect device as the first transistor. If the load resistance R_L satisfies $R_L \gg f_TR/f$ the over-all gain is approximately g_mR, and the second stage is effectively a grounded base stage with an input admittance equal to its mutual conductance. Thus again we obtain (13.36). Now g_m is the mutual conductance of the first stage and is not directly related to I_e. On the other hand, if R_s is optimized its value is near f_T/fg_m and so

$$F \sim F_1 + \frac{eI_e f}{2kT_s g_m f_T}. \tag{13.37}$$

If, for example, we take $T_s = 300$ K, $I_e = 1$ mA, $f = \frac{1}{2}f_T$, and

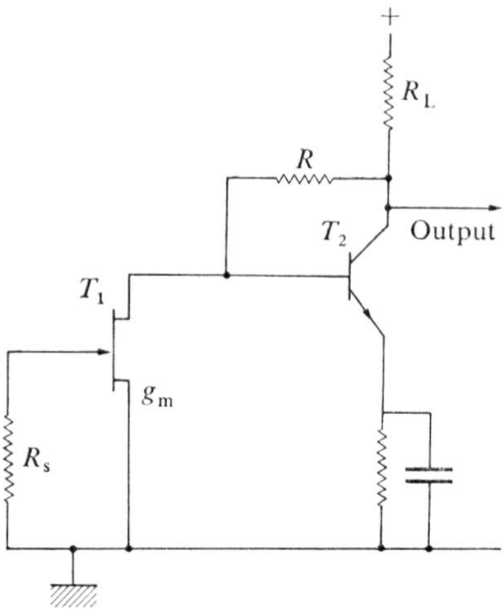

FIG. 13.6. The Cherry and Hooper mismatched pair.

$g_m = 5$ mA V^{-1} we obtain $F = F_1 + 2$, which is a rather unacceptable increase. If on the other hand, the first stage is a bipolar transistor with the same collector current as T_2 we shall have $F \sim F_1$. Thus, even if the field-effect device has $F_1 = 1{\cdot}5$ and the bipolar device has $F_1 = 2$, the bipolar device will yield $F \sim 2$ and the f.e.t. will yield $F \sim 3{\cdot}5$.

13.4. A comparison of devices

At relatively low frequencies the noise temperature of a bipolar device is, at best, $\beta_{dc}^{-\frac{1}{2}}T \sim T/20$ whereas, especially if the source impedance is high, field-effect devices yield $T/200$ or less. Thus in critical applications up to about 1–10 MHz the field-effect device is preferable. However, if the difference between $T/2$ and $T/200$ is insignificant, which it is in most cases, the choice can be dictated by other considerations, e.g. the ease of obtaining the appropriate optimum source impedance or considerations not directly related to noise. One such consideration is the relation between g_m and the standing bias current. Usually an f.e.t. requires 50 times as much current to yield the same value of g_m as a bipolar transistor.

The noise temperature of both devices increases as f approaches f_T, but because it starts at a lower level in field-effect devices the actual magnitudes are comparable at different ratios of f to f_T. Roughly speaking, an f.e.t. at $f \sim f_T/2$ is about as bad as a bipolar transistor at $f_T/10$, but above f_T both devices deteriorate very rapidly. Since field-effect devices with f_T above about 200 MHz are expensive, whereas bipolar devices with f_T up to 2000 MHz are still relatively inexpensive, the bipolar device tends to be more useful above about 100 MHz and is the only useful device above 500 MHz.

Since this was first written Baechtold, Walter, and Wolf (1972) have reported noise figures as low as 5 dB at 10 000 MHz and 2 dB at 2000 MHz obtained with field-effect transistors constructed with gallium arsenide channels and Schottky diode gates. Gallium arsenide has a higher carrier mobility than silicon and so less mobile channel charge is required for the same conductance. The reduction in charge decreases the gate-to-channel capacitance and so increases $f_T = g_m/2\pi C$. At the same time bipolar transistors with $f_T \sim 5000$ MHz have become cheaper and more readily available. Thus in a short time the upper limit to the frequency of operation of both types of transistor has been significantly increased.

Generally vacuum tubes are noisier than either type of transistor at all frequencies, but nevertheless give acceptable performance with $T_n < 300$ K up to about 100 MHz. They have one major advantage: they are resistance to damage by transient overload. In a rugged environment a vacuum-tube amplifier which maintains a noise temperature of 300 K may be preferable to a transistor amplifier which, although it starts with $T_n = 30$ K, soon deteriorates to $T_n \sim 3000$ K.

Finally, we may remark that in many cases a noise temperature of 1000 K may be adequate. It is then a waste of time and effort to produce a refined design with a much lower noise temperature.

14 Low-Frequency Amplifiers

14.1. Field-effect transistors

The noise in the channel current of a junction f.e.t. begins to rise as the frequency decreases, and typically is doubled at 25 kHz. The equivalent circuit of an amplifier stage with a source resistance R_s is shown in Fig. 14.1, and the power spectra of the three sources are

$$w_s = 4kT_sR_s, \tag{14.1a}$$

$$w_i = 2eI_g = 4kTg_n, \tag{14.1b}$$

$$w_v = \frac{8}{3}\frac{kT}{g_m}\left(1+\frac{f_0}{f}\right) = 4kTr_n. \tag{14.1c}$$

The parameter f_0 determines the frequency at which flicker noise begins to be significant.

The optimum source resistance is

$$R_{s0} = \left(\frac{r_n}{g_n}\right)^{\frac{1}{2}} = \left(\frac{4kT}{3eI_g g_m}\right)^{\frac{1}{2}}\left(1+\frac{f_0}{f}\right)^{\frac{1}{2}}, \tag{14.2}$$

and the optimum noise temperature is

$$T_{n0} = 2T(r_n g_n)^{\frac{1}{2}} = T\left(\frac{4eI_g}{3kTg_m}\right)^{\frac{1}{2}}\left(1+\frac{f_0}{f}\right)^{\frac{1}{2}}. \tag{14.3}$$

Representative values of I_g and g_m for low-noise devices are 10^{-11} A and 5 mA V^{-1}, so that

$$R_{s0} \sim 10^6(1+f_0/f)^{\frac{1}{2}}\ \Omega. \tag{14.4}$$

If $f_0 = 25\times10^3$ Hz and $f = 10$ Hz we have $R_{s0} \sim 5\times10^7\ \Omega$. Usually the resistance of the signal source will be less than this and, as we shall see in § 14.3, it will not be possible to transform R_s up to such a high value. Thus we shall have $R_s \ll R_{s0}$ and

$$T_n \sim T\frac{r_n}{R_s} = \frac{2T}{3R_s g_m}\left(1+\frac{f_0}{f}\right). \tag{14.5a}$$

If, for example, $g_m = 5$ mA V^{-1} and $f_0 = 25\times10^3$ Hz we shall have

$$\frac{T_n}{T} \sim \frac{3\times10^6}{R_s f}. \tag{14.5b}$$

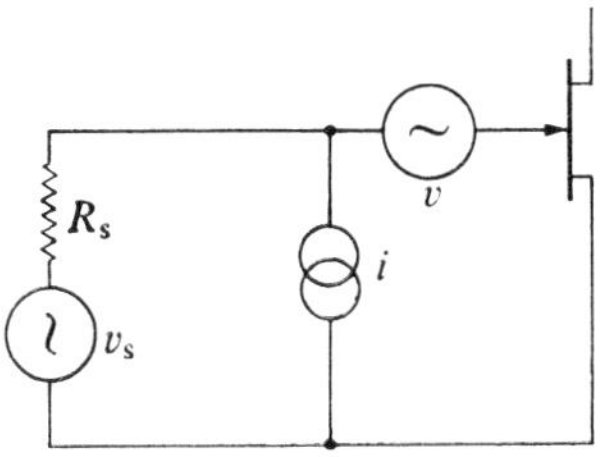

FIG. 14.1. F.e.t. input stage.

Thus with $R_s = 10^6\ \Omega$ and $f = 10$ Hz we get $T_n/T \sim 0{\cdot}3$ and $F = 1{\cdot}3$. If however $R_s = 3000\ \Omega$ we have $T_n/T \sim 100$ and $F \sim 100$. We see that low-noise amplification at low audio-frequencies is feasible if R_s is large but not if R_s is small.

14.2. Bipolar transistors

In these devices the additional low-frequency noise is associated with the base current, and the power spectra of the noise generators in Fig. 14.2 are

$$w_s = 4kT_sR_s \tag{14.6a}$$

$$w_b = 4kTR_b \tag{14.6b}$$

$$w_v = \frac{2kT}{g_m} = 4kTr_n \tag{14.6c}$$

and

$$w_i = \frac{2kTg_m}{\beta_{dc}}\left(1+\frac{f_0}{f}\right) = 4kTg_n. \tag{14.6d}$$

The noise temperature is

$$T_n = T\left\{\frac{R_b}{R_s}+\frac{r_n}{R_s}+\frac{(R_s+R_b)^2}{R_s}g_n\right\}, \tag{14.7}$$

which we can express also as

$$T_n = T\left\{\frac{R_b}{R_s}+\frac{1}{2g_mR_s}+\frac{(R_s+R_b)^2}{2\beta_{dc}R_s}g_m+\frac{(R_s+R_b)^2f_0}{2\beta_{dc}R_sf}g_m\right\}. \tag{14.8}$$

We see that, whatever the bias current and the value of g_m, the excess-noise term is always minimized when $R_s = R_b$. However, if $R_s = R_b$ we also see that $T_n > T$.

The over-all minimum is obtained with

$$R_{s0} = \frac{\beta_{dc}^{\frac{1}{2}}}{g_m}\left(\frac{1+2R_bg_m}{1+f_0/f}+R_b^2g_m^2\beta_{dc}^{-1}\right) \tag{14.9}$$

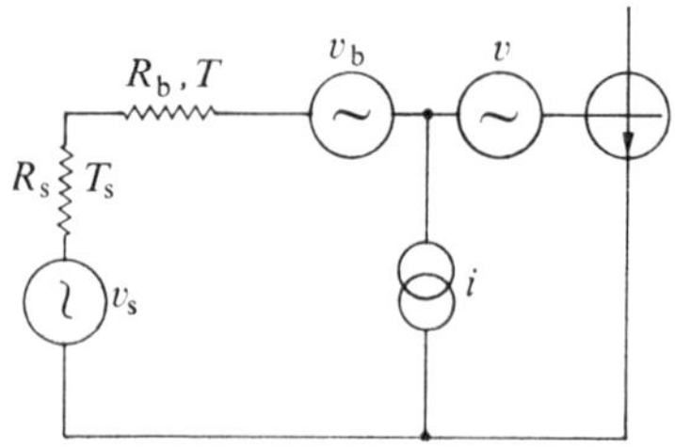

FIG. 14.2. Bipolar transistor input stage.

and then

$$\frac{T_{n0}}{T} = \left(1+\frac{f_0}{f}\right)R_b\frac{g_m}{\beta_{dc}}+\beta_{dc}^{-\frac{1}{2}}\left[\left(1+\frac{f_0}{f}\right)\left\{1+2R_b g_m+\left(1+\frac{f_0}{f}\right)R_b^2 g_m^2 \beta_{dc}^{-1}\right\}\right]^{\frac{1}{2}}. \tag{14.10}$$

We notice that T_{n0} decreases as the collector current I_c and g_m are decreased to the lowest possible value compatible with a high value of β_{dc}. For example, the p–n–p type BCY 71A has $\beta_{dc} \sim 100$ at $I_c = 10\ \mu A$, where $g_m = 400\ \mu A\ V^{-1}$ and since, for this transistor, $R_b = 200\ \Omega$ and $f_0 = 1000$ Hz down to a few hertz we can neglect all the terms involving R_b. This gives

$$R_{s0} \sim \frac{\beta_{dc}^{\frac{1}{2}}}{g_m}\left(1+\frac{f_0}{f}\right)^{-\frac{1}{2}} \sim 2.5\times 10^4\left(1+\frac{1000}{f}\right)^{-\frac{1}{2}}\ \Omega \tag{14.11a}$$

and

$$\frac{T_{n0}}{T} \sim \left(\frac{1+f_0/f}{\beta_{dc}}\right)^{\frac{1}{2}} \sim \frac{1}{10}\left(1+\frac{1000}{f}\right)^{\frac{1}{2}}. \tag{14.11b}$$

Thus at 10 Hz we have $R_{s0} \sim 2500\ \Omega$ and $T_{n0} \sim T$ or $F \sim 2$. An f.e.t. would give $F \sim 100$ with this source impedance.

If R_s is fixed, T_n can be minimized by altering the transistor bias so that

$$g_m = \left(\frac{\beta_{dc}}{1+f_0/f}\right)^{\frac{1}{2}}(R_s+R_b)^{-1}, \tag{14.12}$$

and this yields

$$\frac{T_n}{T} = \frac{R_b}{R_s}+\frac{R_b+R_s}{R_s}\left(\frac{1+f_0/f}{\beta_{dc}}\right)^{\frac{1}{2}}. \tag{14.13}$$

Thus if $R_b = 200\ \Omega$, $R_s = 100\ \Omega$, $\beta_{dc} = 100$, and $f_0/f = 100$, we have $g_m \sim 3$ mA V^{-1}, requiring $I_c = 75\ \mu A$ and $T_n/T = 5$ or $F = 6$.

14.3. Transformers at low frequencies

At moderate and high frequencies we can usually design low-loss transformers to produce any required source impedance but, as we now show, this is difficult at low frequencies. Fig. 14.3 shows a source of internal resistance R_s connected to a transformer whose primary and secondary inductances are L_1 and L_2. The transformer is ideal in the sense that the resistances

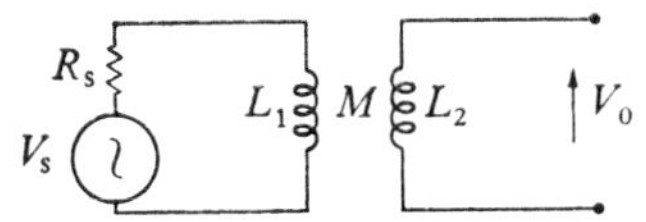

FIG. 14.3. A transformer with a source of internal impedance R_s.

of its windings are negligible and all the primary flux links the secondary so that $M^2 = L_1L_2$. The output impedance is

$$R = R_s\left(\frac{V_0}{V_s}\right)^2 = \frac{R_s\omega^2M^2}{R_s^2+\omega^2L_1^2} = \frac{R_s\omega^2L_1L_2}{R_s^2+\omega^2L_1^2}.$$

This increases as L_2 is increased and we assume that L_2 is already as large as is practicable; then R is maximized if the number of primary turns is chosen so that $\omega L_1 = R_s$, and this gives

$$R = \omega L_2/2. \tag{14.14}$$

Suppose then that the frequency is 10 Hz and we require $R = 3$ MΩ. This requires $L_2 = 10^5$ H, which verges on the limit of practicability. A typical small microphone transformer has $L_2 \sim 300$ H. Thus in designing low-noise, low-frequency amplifiers we will not usually be able to transform the source impedance up to a value where we can take advantage of the inherently low-noise performance of f.e.t.s. Although with $R_s \sim 50$ MΩ an f.e.t. might give $T_n \sim 10$ K at 10 Hz, this is somewhat academic if the source impedance is actually 1 kΩ. Indeed, we should do better to use a bipolar transistor, for eqn (14.13) yields $T_n \sim 1{\cdot}4T \sim 400$ K with this source impedance, whereas (14.5) yields $T_n \sim 300\,T \sim 90\,000$ K.

14.4. Conclusion

Field-effect and bipolar devices are available which make low-noise amplification possible at low frequencies if the source impedance happens either to be very high or in the region of a few thousand ohms, but other components in the amplifier can also generate noise when they are subject to a d.c. bias. Some resistors, especially carbon-composition resistors, are exceedingly noisy, and it is usually essential to avoid these. Manufacturers usually specify the r.m.s. noise generated by a resistor as a fraction of t applied bias voltage, a typical figure is 0·1 μV per volt—this refers to total integrated low-frequency noise. To get some idea of what this means we note that the r.m.s. noise voltage per root-hertz due to Johnson noise in a 1 MΩ resistance at 300 K is also about 0·1 μV.

We might expect electrolytic capacitors, thermistors, and Zener diodes to be noisy, but some rather unexpected components are also surprisingly noisy, e.g. disc ceramic capacitors. Thus in designing low-noise, low-frequency amplifiers all the components, not just the active devices, have to be chosen with care.

15 Electron-beam tubes

15.1. Introduction

ABOVE about 1 GHz low-noise amplification using transistors or conventional electron tubes becomes progressively more difficult, and either parametric amplifiers or masers, to be discussed in the next two chapters, are required for applications in which noise performance is critical. There is, however, a range of applications where a noise figure of 3–10 dB is tolerable, and for these purposes electron-beam tubes, especially travelling-wave tubes and backward-wave tubes, are suitable. They have the further advantage of a large bandwidth (the gain–bandwidth product of a travelling-wave tube usually exceeds 10^5 MHz) and considerable resistance to damage by overloading.

Electron-beam tubes can be divided broadly into two classes: those in which a circuit interacts with longitudinal space-charge waves in the beam and those in which the circuit interacts with transverse cyclotron waves. The travelling-wave tube belongs to the first class and the Adler tube to the second (Adler, Hrbek, and Wade, 1959).

15.2. Gain in longitudinal-wave tubes

The gain mechanism in all longitudinal-wave tubes, whether they are klystrons or travelling-wave tubes, is essentially the same and can be discussed in terms of the two space-charge waves supported by an electron beam of uniform drift velocity v_0. One of these waves travels slightly faster than v_0 and the other slightly slower. The waves are analogous to compression, or sound, waves in a fluid flowing at a supersonic velocity. In the fast space-charge wave the compression and the velocity-modulation amplitude are in phase, and a beam supporting a fast wave has an increased energy. In the slow wave, compression and velocity are out of phase, and the beam has a decreased energy. The signal power transmitted by a beam of mean velocity v_0 and mean current I_0 supporting space-charge waves of small velocity amplitude v_1 and current modulation amplitude I_1 can (Chu 1951) be expressed as $(mv_0/e)v_1I_1$, and v_1 and I_1 can be related, in turn, to the amplitudes a_2 and a_3 of the fast and slow space-charge waves. If these are suitably normalized,

$$I_1 = Z^{-\frac{1}{2}}(a_2+a_3) \qquad (15.1a)$$

and

$$\frac{mv_0}{e}v_1 = Z^{\frac{1}{2}}(a_2 - a_3), \tag{15.1b}$$

where Z depends on the frequency, I_0, and v_0. The mean power is then (using r.m.s. amplitude)

$$P = \text{Re}\,\frac{mv_0}{e}v_1 I_1^* = a_2 a_2^* - a_3 a_3^*. \tag{15.2}$$

The fast wave carries positive power, and the slow wave carries negative power. If the beam drifts freely or is accelerated by static fields, P is constant, but if the beam interacts with a radio-frequency circuit, power may enter or leave the beam and there is a corresponding change in the circuit power.

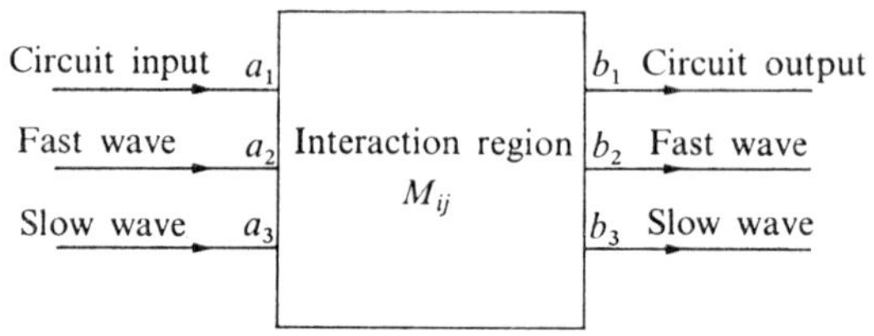

FIG. 15.1. The general form of an electron-beam amplifier.

Fig. (15.1) shows, in its most general form, an electron-beam amplifier. An electron beam with space-charge wave amplitudes a_2 and a_3 together with circuit power $a_1 a_1^*$, of normalized amplitude a_1, enter an interaction region, and the outgoing waves have amplitudes b_1, b_2, and b_3. The circuit power-gain is $b_1 b_1^*/a_1 a_1^*$ and, if this is greater than unity, either the amplitude of the fast wave has decreased or that of the slow wave has increased, for we must have

$$b_1 b_1^* + b_2 b_2^* - b_3 b_3^* = a_1 a_1^* + a_2 a_2^* - a_3 a_3^*. \tag{15.3}$$

The effect of the interaction can be described by a 3×3 matrix with elements M_{ij} so that

$$b_i = \sum_j M_{ij} a_j, \tag{15.4}$$

and, if we regard a and b as column matrices, this can be written in matrix notation as

$$b = Ma \tag{15.5}$$

If we define a^+ and b^+ as the Hermitian conjugates of a and b (i.e. as the transposed matrices with complex conjugate elements) and the diagonal matrix

$$P = \begin{bmatrix} 1 & 0 & 0 \\ 0 & 1 & 0 \\ 0 & 0 & -1 \end{bmatrix},$$

we can express the power conservation law (15.3) as

$$b^{+}Pb = a^{+}Pa.$$

But $b^{+} = a^{+}M^{+}$ and $b = Ma$ and, since a is arbitrary, we must have

$$M^{+}PM = P. \tag{15.6}$$

This places several restrictions on the form of the elements of M. Since the determinant of P does not vanish, M satisfying (15.6) must have a non-vanishing determinant, and therefore an inverse. In addition $P = P^{-1}$, and so we can go through the steps

$$M^{+}P = PM^{-1}, \qquad MPM^{+}P = MPPM^{-1} = MM^{-1} = I, \qquad MPM^{+} = P^{-1}.$$

and obtain

$$MPM^{+} = P. \tag{15.7}$$

The (1, 1) element of this equation is

$$M_{11}M_{11}^{*}+M_{12}M_{12}^{*}-M_{13}M_{13}^{*} = 1.$$

Now $M_{11}M_{11}^{*}$ is the power gain G and so

$$M_{13}M_{13}^{*} = G-1+M_{12}M_{12}^{*}. \tag{15.8}$$

If the gain is large, the coefficient M_{13} which couples noise in the incoming slow wave to the output is also large. Any noise associated with this wave appears amplified at the output. Now noise in the slow wave represents a diminution of the energy of the beam and can only be removed by either adding *correlated* noise power from outside, which is impossible, or by transferring whatever correlated noise power exists in the fast wave to the slow wave. It is this limitation which sets a lower limit to the noise figure of an electron-beam tube.

This description of excitations in an electron beam is only valid if the velocity of the electrons in the beam is a single-valued function of position and time. However, it can be shown to be an exceedingly good approximation, even in a beam in which the electrons have a spread of velocities about the mean value, provided that the mean square velocity spread satisfies $\langle \Delta v^2 \rangle < (\omega_q/\omega)v_0^2$, where ω is the frequency of the wave components under discussion and ω_q is the effective plasma frequency. This parameter measures the strength of the space-charge interactions which reduce the disorder of the electron flow. In a beam from a cathode at T_c, accelerated to a voltage V, the critical condition is $kT_c < (\omega_q/\omega)$ eV and, in the beams used in microwave tubes, this is satisfied for V greater than about 1 V. Thus the wave description is valid from a point just in front of the cathode throughout the rest of the tube. If we treat this point as the point of entry into the interaction region then, since

$$b_1 = M_{11}a_1+M_{12}a_2+M_{13}a_3,$$

the output noise can be expressed in terms of the power spectra w_1, w_2, w_3, and w_{32} of the incoming circuit wave and the beam waves at this point. The power spectrum of b_1 is

$$W = M_{11}M_{11}^* w_1 + M_{12}M_{12}^* w_2 + M_{13}M_{13}^* w_3 + M_{12}M_{13}^* w_{32} + M_{12}^* M_{13} w_{32}^*,$$

and the noise figure is

$$F = 1 + \frac{M_{12}M_{12}^* w_2 + M_{13}M_{13}^* w_3 + M_{12}M_{13}^* w_{32} + M_{12}^* M_{13} w_{32}^*}{G w_1}, \qquad (15.9)$$

in which we have replaced $M_{11}M_{11}^*$ by the power gain G. If we separate all the variables in (15.9) into their moduli and their arguments so that, for example, $M_{12} \to M_{12} \exp j\theta_{12}$ and $w_{32} \to w_{32} \exp j\phi_{32}$, we have

$$G w_1(F-1) = M_{12}^2 w_2 + M_{13}^2 w_3 + 2M_{12}M_{13}w_{32}\cos(\theta_{12} - \theta_{13} + \phi_{32}).$$

There is no restriction on the phase angles of M_{12} and M_{13} so we can choose them (by suitable changes in the circuit) so that $\theta_{12} - \theta_{13} + \phi_{32} = \pi$, and we then have

$$G w_1(F-1) = M_{12}^2 w_2 + M_{13}^2 w_3 - 2M_{12}M_{13}w_{32}. \qquad (15.10)$$

We are interested in the minimum value of this expression (for given values of G and the power spectra) when M_{12} and M_{13} are regarded as variables limited only by the relation (15.8) which now becomes

$$M_{13}^2 = G - 1 + M_{12}^2.$$

This result is expressed most conveniently in terms of the noise parameters

$$S = \{(w_2 + w_3)^2 - 4w_{32}^2\}^{\frac{1}{2}} \qquad (15.11)$$

and $\Pi = w_2 - w_3$ and is

$$F - 1 = \frac{G-1}{G} \cdot \frac{S - \Pi}{2w_1}. \qquad (15.12)$$

We see that both F and the noise measure

$$M = \frac{F-1}{1-1/G} = \frac{S-\Pi}{2w_1} \qquad (15.13)$$

are fixed by the noise content $(S - \Pi)$ of the beam at the point where space-charge wave propagation begins.

From eqns (15.1a) and (15.1b) we have

$$a_2 = \frac{1}{2}\left(Z^{\frac{1}{2}} I_1 + \frac{m v_0}{e} Z^{-\frac{1}{2}} v_1\right), \qquad a_3 = \frac{1}{2}\left(Z^{\frac{1}{2}} I_1 - \frac{m v_0}{e} Z^{-\frac{1}{2}} v_1\right),$$

and we can use these relations to express w_2, w_3, and w_{32}, or S and π, in terms of the power spectra of I_1 and v_1. The result is

$$S = \frac{mv_0}{e}\left\{4w_i w_v+(w_{iv}-w_{iv}^*)^2\right\}^{\frac{1}{2}}, \qquad \Pi = \frac{mv_0}{2e}(w_{iv}+w_{iv}^*). \quad (15.14)$$

The values of w_i, w_v, and w_{iv} now have to be evaluated at a point near the cathode. We consider this problem in the next section.

15.3. The noise content of an electron beam

Conditions in the electron flow just in front of the cathode are exceedingly complex. The electron velocity is not a single-valued function of time and position, the Maxwellian distribution of thermal velocities is significant, and space–charge interactions are strong. Thus to evaluate w_i, w_v, and w_{iv} we shall have to make some very drastic approximations. There are essentially two extreme approaches possible. In the first we assume that space-charge interaction in this cathode region is negligible and that the electrons arrive independently and randomly at the input plane of the interaction region. In the second we assume that the conditions in the cathode region are quasi-static so that we can use the Thompson–North–Harris (1940) treatment of space-charge smoothing.

With the first assumption we can use the results obtained in § 5.4. At a plane where the potential is $V = \alpha kT_c/e$ relative to the cathode we have

$$w_i = 2eI_0, \qquad w_{iv} = 0, \qquad \text{and} \qquad w_v = \frac{4ekT_c}{mI_0}f(\alpha), \quad (15.15)$$

so that $\Pi = 0$ and

$$S = \frac{mv_0}{e}(w_i w_v) = \{2mv_0^2 kT_c . 4f(\alpha)\}^{\frac{1}{2}} = 2kT_c\{4(1+\alpha)f(\alpha)\}^{\frac{1}{2}}. \quad (15.16)$$

The function $4(1+\alpha)f(\alpha)$ is extremely insensitive to the value of α and increases steadily from $0{\cdot}86 = 4-\pi$ to unity as α increases from zero to infinity. To a very good approximation $S = 2kT_c$ and so, since $w_1 = kT$, where T is the source temperature, the noise figure at high gain is

$$F \sim 1+T_c/T. \quad (15.17a)$$

and the noise temperature is

$$T_n \sim T_c. \quad (15.17b)$$

In the other extreme case the current fluctuations are greatly reduced by space–charge smoothing and we have (Robinson, 1958b)

$$w_i = \Gamma^2 2eI_0 = 9\frac{f(0)}{1+\alpha}2eI_0.$$

The value of w_v is still given by (14.15) but now, since increased current is associated with an increased emission of energetic electrons, the current and voltage fluctuations are correlated almost completely, thus so $w_{iv} = (w_i w_v^*)^{\frac{1}{2}}$ and is real. As a result

$$S = \Pi = \frac{mv_0}{e}(w_i w_v)^{\frac{1}{2}} = 12kT_c\{f(\alpha)f(0)\}^{\frac{1}{2}}. \quad (15.18)$$

We note that, if $\alpha = eV/kT_c$ is not large, the value of S is much the same as (15.16) but the noise content ($S-\Pi$) is almost zero. Thus the contribution of the electron-beam noise to the tube is also almost zero. We shall discuss the relevance of these results to practical tubes in a later section.

15.4. The Adler tube

Although the Adler tube has not found any widespread applications it is of some interest since, in principle at least, it offers the possibility of obtaining almost noiseless amplification in an electron-beam tube. In this device, shown schematically in Fig. 15.2, space-charge in the low-current, low-voltage electron beam is almost negligible, but there is an axial magnetic field **B** parallel to the beam, arranged so that the signal frequency f coincides with the cyclotron frequency $f_c = eB/2\pi m$. The signal is applied between the

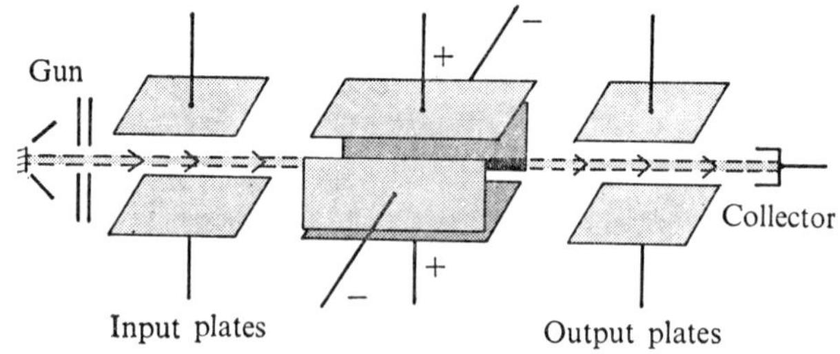

FIG. 15.2. Adler tube.

first pair of plates and, if their length and the beam current are correctly chosen, all the input signal power is delivered to the beam and causes the electrons to spiral coherently about the axis. The beam then enters the amplifying region where a quadupole arrangement of electrodes is driven by an external oscillator at $2f_c$. In this region the amplitude of the spiralling motion is increased but remains proportional to the initial motion impressed by the input plates. The beam emerging from the quadrupole region then delivers its energy to the output plates. Since all the active circuits couple only to transverse motion, fluctuations in longitudinal current and velocity lead to no output noise. The noise temperature is determined solely by the fluctuations in the transverse velocity of the beam where it enters the input plates, and it is not difficult to see that the effective noise temperature will be approximately $mv^2/2k$, where v^2 is the mean square spread in transverse velocity.

This is determined not only by the temperature of the cathode but also by the angular aperture of the collimating system used in the gun. If the half-angle is θ we have in terms of the longitudinal beam velocity v_0, for small values of θ, $v^2 \sim \frac{1}{3}\theta^2 v_0^2$. Thus $kT_n \sim \frac{2}{3}\theta^2 eV_0$, where V_0 is the d.c. beam voltage. If $V_0 = 6$ V and $\theta = \frac{1}{20}$ rad (2·8°) this gives $T_n \sim 100$ K. Further collimation of the electron beam reduces T_n, and a practical limit is set only by the need to get a reasonable (a few micro-amps) beam current. Unfortunately, this requirement is not trivial, and as a result Adler tubes are rather difficult to construct, especially at frequencies above 1 GHz. They also require pump power at $2f$, and if this is available it is more economical to use it to drive a parametric amplifier. However, the tube does have one practical advantage. It is not only immune to overload but it also acts as a power limiter and so protects succeeding amplifiers from damage by overload.

15.5. Travelling-wave tubes

Although the considerations discussed in § 5.3 apply to all longitudinal electron-beam tubes, in practice only travelling-wave tubes and their close relation the backward-wave tube can be built with noise figures approaching the theoretical limit. The reason for this is essentially that in these two tubes the interaction between the beam and the circuit is strong. This makes it feasible to obtain a large gain at very low beam currents, of the order of a few hundred micro-amperes. Now, although the equations that we have derived set a lower limit to the noise performance attainable with a beam from a thermionic cathode, they do not take into account any deterioration in the properties of the beam caused by poor electron-optics. Any process which either intercepts a fraction of the beam or introduces an additional spread in the *axial* velocity component of the electrons greatly increases the noise content of the beam. These effects are easily avoided only with low-current beams, which can be formed with very simple electron-optical systems and so the ultimate noise performance is only approached by low-current devices such as the travelling-wave tube.

The approach to the calculation of the noise content of the beam which neglects space-charge interaction near the cathode leads to $S \sim 2kT_c$, $\Pi = 0$, and $T_n \sim T_c$. If $T_c \sim 900$ K this corresponds to a noise figure of 4 (or 6 dB) relative to a source at 293 K and most simple low-noise travelling-wave tubes indeed have noise figures of this order. Furthermore, direct measurements of beam noise confirm that $S \sim 2kT_c$. However, it is possible to design tubes with lower values of F, down to about $F \sim 2{\cdot}5$ corresponding to $T_n \sim 450$ K. At this level noise generated by ohmic loss in the circuit cannot be neglected, and the contribution to T_n which arises from beam noise is certainly less than 450 K, more probably 300 K $\sim T_c/3$. It is apparent that in these tubes space-charge smoothing near the cathode plays a significant role. This is confirmed in two ways. First, it is found that S measured directly is not much changed in these very low-noise electron beams. Thus the improvement in $kT_n = \frac{1}{2}(S - \Pi)$ must come from an increase in Π. As we have seen, this is expected when space-charge smoothing is effective. Secondly, the devices in which low-noise performance is achieved all work with an electron beam derived from an extremely small cathode area. This means that the current density at the cathode is increased, and this increases the plasma frequency ω_q in this region. Since the transition from quasi-static behaviour, displaying space-charge smoothing, to the true high-frequency regime, where smoothing is inoperative, depends on the ratio ω/ω_q, an increase in ω_q tends to make smoothing processes more effective at any given frequency.

Travelling-wave tubes with noise figures as low as 3·5 dB are available in the 1–3 GHz region, and in this region tubes with $F \sim 7$ dB are also available at much less cost. The best value of F increases with frequency, 10 dB is about typical at 20 GHz. These figures, corresponding to noise temperatures of hundreds or thousands of kelvins, do not compare with those attainable with parametric amplifiers or, if refrigeration to 4 K is not obstacle, with masers. They are adequate, however, for many purposes. We may remark also that travelling-wave tubes are not only electrically robust but they also have quite remarkable bandwidths. Whereas with any other device a voltage gain of 10 (power gain of 20 dB) over a bandwidth of 100 MHz is a considerable achievement, it is difficult to design a travelling-wave tube of the same gain to have less bandwidth. A bandwidth of ± 1000 MHz centred at 4000 MHz is a much more typical figure. If we compare this gain–bandwidth product 20 000 MHz with the values around 2000 MHz typical of good bipolar transistors or 200 MHz typical of field-effect devices and vacuum triodes we see the spectacular difference between travelling-wave tubes and any other amplifying device.

16 Parametric Amplifiers

16.1. Introduction

A CIRCUIT containing a non-linear reactive element excited at several frequencies simultaneously will generate combination frequencies, and it is possible, in certain circumstances, for an excitation at a frequency ω_1 to lead to the generation of more power at ω_1 than is provided by the external source of excitation. The system is then an amplifier. Parametric processes in mechanical systems were known to Rayleigh and, although not recognized as potential sources of amplification, have been known for many years to play a significant role in some solid-state phenomena. The first suggestion however that they might be of use as an amplification mechanism was due to Suhl (1957), who considered parametric processes in ferromagnetic resonance. This particular type of parametric process has not led to useful practical devices, and parametric amplifiers in current use are almost exclusively based on the use of the non-linear voltage–charge relationship found in a reverse-biased p–n junction. Although, in practical amplifiers, a number of different circuit configurations are used, they are all derived from the basic configuration shown in Fig. 16.1, in which a diode is connected to external sources by series-tuned circuits which only pass frequencies ω_1, ω_2, and $\omega_1+\omega_2$. If the diode voltage–charge relation is

$$V = \alpha q+\beta q^2.$$

and the impressed charge is

$$q = \tfrac{1}{2}[Q_1 \exp(j\omega_1 t)+Q_1^* \exp(-j\omega_1 t)+Q_2 \exp(j\omega_2 t)+Q_2^* \exp(-j\omega_2 t)+ \\ +Q_{12} \exp\{j(\omega_1+\omega_2)t\}+Q_{12}^* \exp\{-j(\omega_1+\omega_2)t\}],$$

the diode current is

$$I = \dot{q} = \tfrac{1}{2}\{j\omega_1 Q_1 \exp(j\omega_1 t)-j\omega_1 Q_1^* \exp(-j\omega_1 t)+\text{etc.}\}$$

and the voltage across the diode contains, in addition to linear terms such as $\tfrac{1}{2}\alpha Q_1 \exp(j\omega_1 t)$, a group of non-linear terms at the frequencies ω_1, ω_2, and $\omega_1+\omega_2$. These give

$$V = \tfrac{1}{2}\beta[Q_1 Q_2 \exp\{j(\omega_1+\omega_2)t\}+\text{c.c.}+Q_{12}Q_1^* \exp(j\omega_2 t)+\text{c.c.}+ \\ +Q_{12}Q_2^* \exp(j\omega_1 t)+\text{c.c.}],$$

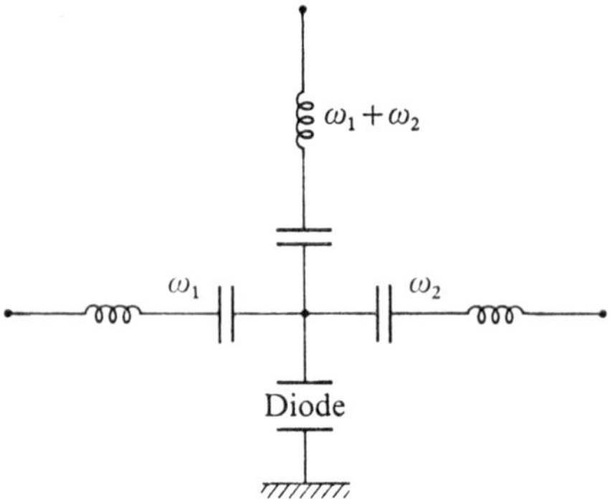

FIG. 16.1. Parametric amplifier using a diode as a non-linear capacitance.

where c.c. stands for complex conjugate. The time-average power delivered by the external source at ω_1 to the diode is therefore

$$P_1 = \tfrac{1}{4}\mathrm{j}\omega_1\beta(Q_1Q_2Q_{12}^*-Q_1^*Q_2^*Q_{12}),$$

and similarly the powers delivered to the diode at ω_2 and $\omega_1+\omega_2$ are

$$P_2 = \tfrac{1}{4}\mathrm{j}\omega_2\beta(Q_1Q_2Q_{12}^*-Q_1^*Q_2^*Q_{12})$$

and

$$P_{12} = \tfrac{1}{4}\mathrm{j}(\omega_1+\omega_2)\beta(Q_1^*Q_2^*Q_{12}-Q_1Q_2Q_{12}^*).$$

We note first that $P_1+P_2+P_{12} = 0$, so that energy is conserved, but we also have

$$\frac{P_1}{\omega_1} = \frac{P_2}{\omega_2} = -\frac{P_{12}}{\omega_1+\omega_2}. \tag{16.1}$$

This relation was first derived by Manley and Rowe (1956) and is generally known as a Manley–Rowe relation. If the device is to act as an amplifier at ω_1 it must generate more power than it receives at ω_1, i.e. P_1 must be negative. This implies that P_2 is also negative and P_{12} is positive. If the circuit is excited externally by a pump oscillator at $\omega_1+\omega_2$, the power generated at the idler frequency ω_2 must be dissipated in a load, i.e. a resistive element. Thus, although at first sight it appears that only a reactive, and potentially noiseless, element is required, in fact a resistive, and therefore noisy, idler load is also required. As a result, parametric amplifiers, contrary to early expectations, are not noiseless. As we shall see, the minimum possible noise temperature as an amplifier at ω_1 is determined by the temperature T_i of the idler load and is

$$T_\mathrm{n} = \frac{\omega_1}{\omega_2}T_\mathrm{i}. \tag{16.2}$$

In practice, with refrigerated diodes and circuits, values of T_n of the order of a few tens of degrees Kelvin can be achieved at microwave frequencies.

The parametric amplifier is therefore significantly less noisy than a travelling-wave tube, though not as good as a maser.

We now look in more detail at the gain mechanism, which clearly arises from the presence of the pump excitation, and it will be convenient to eliminate this from explicit consideration and to treat its effect as a modulation p, arising from the non-linear term β, of the diode reactance. We therefore let C be the mean diode capacitance and write the differential reactance for small signals at ω_1 and ω_2 as

$$\frac{\mathrm{d}V}{\mathrm{d}q} = \frac{1}{C}[1+p\exp\{\mathrm{j}(\omega_1+\omega_2)t\}+p^*\exp\{-\mathrm{j}(\omega_1+\omega_2)t\}]. \tag{16.3}$$

With the small signal components of the diode voltage, charge, and current expressed as

$$v = \tfrac{1}{2}\{V_1\exp(\mathrm{j}\omega_1 t)+V_1^*\exp(-\mathrm{j}\omega_1 t)+V_2\exp(\mathrm{j}\omega_2 t)+V_2^*\exp(-\mathrm{j}\omega_2 t)\}.$$

$$q = \tfrac{1}{2}\{Q_1\exp(\mathrm{j}\omega_1 t)+Q_1^*\exp(-\mathrm{j}\omega_1 t)+Q_2\exp(\mathrm{j}\omega_2 t)+Q_2^*\exp(-\mathrm{j}\omega_2 t)\},$$

$$I = \dot{q} = \tfrac{1}{2}\{I_1\exp(\mathrm{j}\omega_1 t)+I_1^*\exp(-\mathrm{j}\omega_1 t)+I_2\exp(\mathrm{j}\omega_2 t)+I_2^*\exp(-\mathrm{j}\omega_2 t)\},$$

eqn (16.3) yields the basic small-signal relations

$$V_1 = I_1/\mathrm{j}\omega_1 C - pI_2^*/\mathrm{j}\omega_2 C, \tag{16.4a}$$

$$V_2^* = -I_2/\mathrm{j}\omega_2 C + p^*I_1/\mathrm{j}\omega_1 C. \tag{16.4b}$$

In Fig. 16.2 the inductance L_1 tunes with the mean diode capacitance C at ω_1 and the inductance L_2 at ω_2. If the input signal voltage v_s at ω_1 is expressed as

$$v_s = \tfrac{1}{2}\{V_s\exp(\mathrm{j}\omega_1 t)+V_s^*\exp(-\mathrm{j}\omega_1 t)\}, \tag{16.5}$$

the circuit equations for the signal and idler loops are

$$V_s = R_1 I_1 + \mathrm{j}\omega_1 L_1 I_1 + V_1 = R_1 I_1 - \frac{1}{\mathrm{j}\omega_1 C}I_1 + V_1$$

and

$$0 = R_2 I_2^* + \frac{1}{\mathrm{j}\omega_2 C}I_2^* + V_2^*.$$

Together with (16.4a) and (16.4b) these equations yield

$$I_1 = \frac{\omega_1\omega_2 C^2 R_2 V_s}{\omega_1\omega_2 C^2 R_1 R_2 - pp^*} \tag{16.6a}$$

and

$$I_2^* = \frac{\mathrm{j}\omega_2 C p^* V_s}{\omega_1\omega_2 C^2 R_1 R_2 - pp^*}. \tag{16.6b}$$

We now suppose that $R_1 = r_s + r_l$ where r_s is the source impedance and r_l is a load impedance. The power delivered to the load (we use r.m.s. amplitudes) is $I_1 I_1^* r_l$ and the available power from the source is $V_s V_s^*/4r_s$, so that the power gain is

$$G = \frac{4\omega_1^2\omega_2^2R_2^2r_s r_l C^4}{(\omega_1\omega_2R_1R_2C^2 - pp^*)^2}. \tag{16.7}$$

The system will be stable if $pp^* < \omega_1\omega_2R_1R_2C^2$ and the gain will be large if

$$pp^* \sim \omega_1\omega_2R_1R_2C^2. \tag{16.8}$$

Since $R_1 = r_s + r_l$, if the high-gain condition (16.8) can be fulfilled for some finite value of r_l it can also be fulfilled for a lower value of r_l, and we can

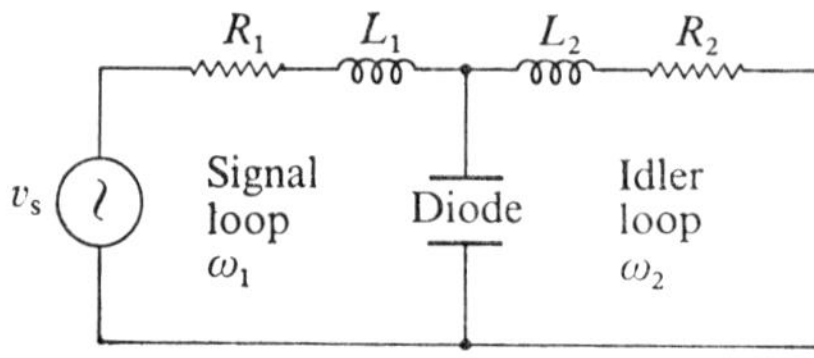

FIG. 16.2. Simplified parametric amplifier circuit.

get high gain for arbitrarily low values of r_l. When we come to consider the noise performance this will allow us to neglect any Johnson-noise voltage generated in the load r_l.

It is clear from (16.7) that the gain depends not only on the circuit parameters R_1, R_2, C etc. but also on the parameter p, which is essentially the fractional change in the diode reactance induced by the pump excitation. For a given set of circuit parameters increasing values of pump excitation will increase the gain until, at some critical value, the circuit breaks into oscillation.

In the next section we shall consider the noise performance of the amplifier shown in Fig. 16.2, and our treatment will be given in elementary classical terms; the reader who is interested in the quantum limitations on parametric amplifier performance and the significance of parametric processes in other contexts should consult Louisell (1964), and the reader who requires a more general treatment of parametric amplifier circuits will find Penfield and Rafuse (1962) a useful text.

16.2. Noise

The fundamental and inescapable noise sources in a parametric amplifier are the source impedance and the idler load which display Johnson noise, but in practice there is also an important contribution from the series resistance r_d of the diode structure. The effect of noise in the signal load, as we have seen, can be neglected, although clearly this implies that, in any practical design, steps have to be taken to ensure that this noise is indeed negligible.

In Fig. 16.3 we show the basic elements of the amplifier together with the noise generators v_n, v_i, and v_d associated with the source resistance r_s at T_s, the idler resistance r_i at T_i, and the diode resistance r_d at T_d. These three noise sources are uncorrelated and, in addition, components of v_d at ω_1 which excite the signal circuit are uncorrelated with components of v_d at ω_2 which excite the idler circuit. For simplicity, in Fig. 16.3 we have omitted the inductances, but it is plain in any case that the total resistance R_1 in the signal circuit is r_s+r_d and in the idler circuit $R_2 = r_i+r_d$, while at the same time

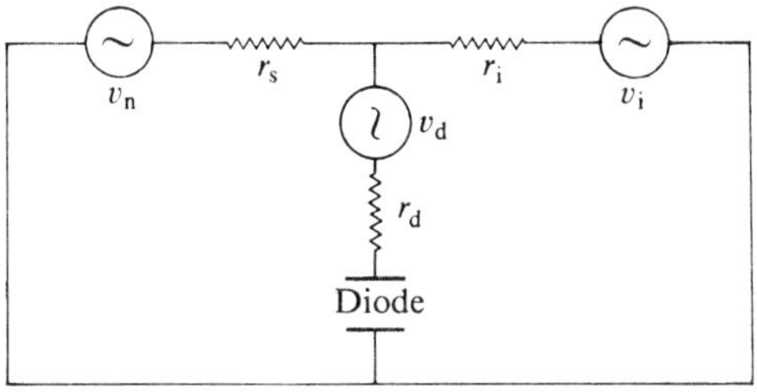

FIG. 16.3. Parametric amplifier with noise sources.

the noise sources in the two circuits are v_n and v_d in the signal circuit and v_i and v_d in the idler circuit. It is also clear from the symmetry of the system that, just as eqn (16.6b) gives I_2^* in terms of the voltage in the signal circuit, we also have

$$I_1 = \frac{-j\omega_1 CpV_2^*}{\omega_1\omega_2 C^2 R_1 R_2 - pp^*}$$

in terms of the voltage in the idler circuit. Thus setting $dV_n^2 = 4kT_s r_s\, df$, etc. we obtain for I_1

$$dI_1^2 = \left(\frac{\omega_1\omega_2 C^2 R_2}{\omega_1\omega_2 C^2 R_1 R_2 - pp^*}\right)^2 (4kT_s r_d\, df + 4kT_d r_d\, df) +$$

$$+ \frac{\omega_1^2 C^2 pp^*}{(\omega_1\omega_2 C^2 R_1 R_2 - pp^*)^2}(4kT_i r_i\, df + 4kT_d r_d\, df).$$

The noise figure F is obtained by dividing this expression by the noise due to r_s alone and is

$$F = 1 + \frac{r_d T_d}{r_s T_s}\left(1 + \frac{pp^*}{\omega_2^2 C^2 R_2^2}\right) + \frac{r_i T_i}{r_s T_s}\left(\frac{pp^*}{\omega_2^2 C^2 R_2^2}\right), \tag{16.9}$$

so that the noise temperature is

$$T_n = T_d \frac{r_d}{r_s}\left(1 + \frac{pp^*}{\omega_2^2 C^2 R_2^2}\right) + T_i \frac{r_i}{r_s}\left(\frac{pp^*}{\omega_2^2 C^2 R_2^2}\right). \tag{16.10}$$

In practice the diode and idler temperatures will be the same, and if we note that $R_2 = r_i + r_d$ this leads to

$$T_n = T_d \frac{r_d}{r_s}\left\{1 + \frac{pp^*}{\omega_2^2 C^2 r_d(r_i + r_d)}\right\}. \tag{16.11}$$

At first sight it appears that the noise temperature can be reduced indefinitely by increasing r_s, but we must remember that, if the amplifier is to have appreciable gain,

$$\omega_1\omega_2 C^2(r_s + r_d)(r_i + r_d) \sim pp^*, \tag{16.12}$$

so that there is a limit to the maximum value of r_s, since r_d, C, and the maximum value of pp^* are fixed by the diode structure. If we use (16.12) and (16.11) to eliminate all the variables except ω_1, ω_2, r_s and r_d we obtain

$$T_n = T_d \frac{r_d}{r_s}\left\{1 + \frac{\omega_1}{\omega_2}\frac{r_s + r_d}{r_d}\right\} = T_d\left\{\frac{\omega_1}{\omega_2} + \frac{r_d}{r_s}\left(1 + \frac{\omega_1}{\omega_2}\right)\right\}. \tag{16.13}$$

If we can make $r_s \gg r_d$ this reduces to $T_n = (\omega_1/\omega_2)T_d$, and for given values of ω_1 and ω_2 this is the lowest possible noise temperature. We see from (16.12) that this is most easily achieved when ω_1 and ω_2 are small. If we define a characteristic frequency $\omega_0/2\pi$ for the diode in terms of r_d, its mean capacitance C, and the largest possible modulation of the reactance p by

$$\omega_0^2 C^2 r_d^2 = pp^*, \tag{16.14}$$

eqn (16.12) becomes

$$\frac{\omega_1\omega_2}{\omega_0^2}\left(1 + \frac{r_s}{r_d}\right)\left(1 + \frac{r_i}{r_d}\right) = 1, \tag{16.15}$$

and clearly we can only obtain appreciable gain and yet make $r_s \gg r_d$ if $\omega_1\omega_2 \ll \omega_0^2$. Thus the parameter $\omega_0/2\pi$ sets a limit to the frequency at which the noise temperature $(\omega_1/\omega_2)T_d$ is attainable. If $\omega_1\omega_2$ is much less than ω_0^2 we can also make r_i large compared with r_d, and then eqn (16.10) is approximately

$$T_n \sim T_i \frac{pp^*}{\omega_2^2 r_s r_i C^2} \sim T_i \frac{\omega_1}{\omega_2}. \tag{16.16}$$

In this case, even if the diode temperature is somewhat higher than the idler temperature T_i, we can achieve the limiting noise temperature characteristic of the idler load and ω_1/ω_2. This is not of much practical interest, since at low frequencies much better amplifiers are available, and therefore parametric diodes are only used at very high frequencies, where r_d cannot be neglected.

To discover the high-frequency limitations on T_n it is only necessary to observe that T_n given by (16.13) for a fixed ω_2 decreases with increasing r_s and that r_s given by (16.15) is a maximum when $r_i = 0$, so that the diode loss serves as the idler load. We then have

$$\frac{r_s}{r_d} = \frac{\omega_0^2 - \omega_1\omega_2}{\omega_1\omega_2} \tag{16.17}$$

and

$$T_n = T_d \frac{\omega_1}{\omega_0} \frac{\omega_0/\omega_2 + \omega_2/\omega_0}{1 - \omega_1\omega_2/\omega_0^2}. \tag{16.18}$$

The optimum value of ω_2 is given by

$$\omega_2 = (\omega_0^2 + \omega_1^2)^{\frac{1}{2}} - \omega_1, \tag{16.19}$$

and the over-all minimum of T_n is

$$T_n = 2\frac{\omega_1}{\omega_2} T_d = \frac{2\omega_1 T_d}{(\omega_0^2 + \omega_1^2)^{\frac{1}{2}} - \omega_1}. \tag{16.20}$$

We note that $T_n = T_d$ when $\omega_1 = \omega_0/\sqrt{8}$ and $\omega_2 = \omega_0/\sqrt{2}$, and we might regard this as the upper frequency limit to the useful operation of the diode.

If in eqn (16.3) we take p to be real, we see that the minimum and maximum values of the diode capacitance satisfy

$$\frac{1}{C_{min}} - \frac{1}{C_{max}} = 4p/C.$$

Manufacturers usually specify a high-frequency parameter

$$f_c' = \frac{1}{r_d}\left(\frac{1}{C_{min}} - \frac{1}{C_{max}}\right),$$

and we see this is related to our parameter ω_0 by

$$f_c' = 4\omega_0/2\pi. \tag{16.21}$$

Thus a diode with $f_c' = 100$ GHz will give useful low-noise gain with $T_n \leqslant T_d$ up to about $f_c'/12 \sim 9$ GHz.

Because, as we have remarked, better amplifiers, e.g. f.e.t.s or bipolar transistors, are available at low frequencies, parametric amplifiers generally are used only at high frequencies above about 500 MHz. There they have to compete with travelling-wave tubes with T_n of the order of 500–2000 K and masers with T_n of the order of 1–10 K. As an amplifying device the travelling-wave tube with its large bandwidth and greater gain is vastly superior to either a parametric amplifier or maser, and its cost is comparable with a parametric amplifier, but much less than a maser. It is also, compared with a parametric amplifier, much less susceptible to damage through overloading. Thus, although a parametric amplifier with a refrigerated diode can give a noise temperature of a few tens of degrees Kelvin in the microwave region, the niche it fills in electronic technology is rather limited. A typical example occurs in receivers for satellite communications where the first amplifier stage is a maser with T_n perhaps as low as 3 K and a gain of 20 dB (100). If the over-all receiver noise temperature is not to be dominated by second stage noise the next stage must have $T_n < 300$ K. A parametric amplifier is the obvious candidate; it is simpler than a second maser and has less noise than a travelling-wave tube. Furthermore, since the maser already requires refrigeration, the need to refrigerate the diode is not a serious practical consideration.

We have discussed parametric amplifiers entirely in terms of radio-frequency varactor-diode amplifiers, but we should also remark that various non-linear processes in the optical and infrared response of some solid media lead to parametric gain processes at these frequencies. Here quantum effects are significant and, although practical difficulties may intervene to prevent its realization, the smallest signal power that gives an output of unity signal to noise ratio in a bandwidth $\mathrm{d}f$ at f is

$$\mathrm{d}P = \left\{hf + \frac{hf}{\exp(hf/kT_\mathrm{i}) - 1}\right\} \mathrm{d}f,$$

where T_i is the temperature of the equivalent idler load.

17 Masers

ALTHOUGH practical maser amplifiers can take many forms we shall only consider the rather basic system illustrated in Fig. 17.1. This consists of a signal source, with a resistive internal impedance at a temperature T, matched to a line which at its other end terminates in a structure coupling it to a system of atoms, each of which has a pair of energy levels separated by $h\nu$ and with a finite width. Power incident on the active element is reflected with an increased amplitude, and the reflected wave constitutes the output of the device. In a practical device a non-reciprocal element such as a circulator would be required to separate P_r from P_i, but we ignore this complication.

Although obviously we must treat the active system quantum mechanically we can treat signals classically, as we saw in Chapter 6. If the incident signal power at a frequency ν_1 is P_{si} it will induce transitions between the two levels with equal probability in either direction. We denote the probability per unit time as

$$w = AP_{si}, \tag{17.1}$$

where A is a constant (which may depend on ν). If there are n_+ atoms in the upper level and n_- in the lower level the power extracted from the active system is $h\nu w(n_+-n_-) = h\nu A(n_+-n_-)P_{si}$, and so the reflected signal power is

$$P_{sr} = P_{si}+h\nu A(n_+-n_-)P_{si}, \tag{17.2}$$

and the power gain is

$$G = 1+h\nu A(n_+-n_-). \tag{17.3}$$

The gain is greater than unity if the level population is inverted so that $n_+ > n_-$.

We also saw in Chapter 6 that the noise power incident on the maser from the source in a frequency range $d\nu$ is

$$dP_{ni} = \left\{\frac{h\nu}{\exp(h\nu/kT)-1}+\tfrac{1}{2}h\nu\right\} d\nu. \tag{17.4}$$

The zero-point fluctuations ($\frac{1}{2}h\nu$) cannot be absorbed and can only induce downward transitions. (This loose statement allows us to evade a detailed quantum-mechanical discussion which leads to the same result.) The rate of upward transitions from the lower state to the higher state is therefore

$$r_u = n_-A(dP_{ni}-\tfrac{1}{2}h\nu\, d\nu). \tag{17.5}$$

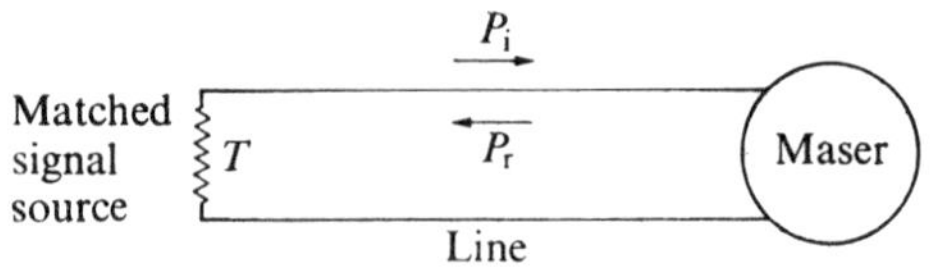

FIG. 17.1. Basic maser configuration.

We can obtain the downward rate r_d without a detailed calculation by considering the situation when the source and the maser medium are in thermal equilibrium at the same temperature. In this case we must have $r_d = r_u$, but we also have $n_+ = n_- \exp(-h\nu/kT)$ and dP_{ni} given by (17.4). We find, as a result, that

$$r_d = n_+ A(dP_{ni} + \tfrac{1}{2}h\nu \, d\nu). \tag{17.6}$$

With the maser levels inverted, so that we no longer have a state of thermal equilibrium, the output noise power is

$$dP_{nr} = dP_{ni} + h\nu(r_d - r_u) = dP_{ni}\{1 + h\nu A(n_+ - n_-)\} + \tfrac{1}{2}h^2\nu^2 A(n_+ + n_-) \, d\nu. \tag{17.7}$$

We can assume that this is so large (it is greater than $G \, dP_{ni}$) that it can be regarded as classically observable. In other words noise in subsequent amplifiers can be neglected. Thus the input signal power for unity signal to noise ratio in the output is

$$P_{si} = dP_{ni} + \tfrac{1}{2}h^2\nu^2 \, d\nu \frac{A}{G}(n_+ + n_-) = dP_{ni} + \left(1 - \frac{1}{G}\right)\frac{n_+ + n_-}{n_+ - n_-}\tfrac{1}{2}h\nu \, d\nu. \tag{17.8}$$

We can define a negative temperature $-T_m$ for the active maser atoms so that

$$\frac{n_+}{n_-} = \exp\left(\frac{h\nu}{kT_m}\right), \tag{17.9}$$

and then we can express P_{si} as

$$P_{si} = \left\{\left(\frac{h\nu}{2} + \frac{h\nu}{\exp(h\nu/kT) - 1}\right) + \left(1 - \frac{1}{G}\right)\left(\frac{h\nu}{2} + \frac{h\nu}{\exp(h\nu/kT_m) - 1}\right)\right\}. \tag{17.10}$$

In the classical limit (low frequencies) this reduces to

$$P_s = \left\{kT + \left(1 - \frac{1}{G}\right)kT_m\right\} d\nu. \tag{17.11}$$

The noise figure is

$$F = 1 + \left(1 - \frac{1}{G}\right)\frac{T_m}{T}, \tag{17.12}$$

and the noise temperature is

$$T_n = \left(1 - \frac{1}{G}\right)T_m. \tag{17.13}$$

If we have a cascade of N identical amplifier stages of noise figure F and gain G the over-all noise figure is

$$F' = F + \frac{F}{G} + \dots \frac{F}{G^{n-1}} = F\frac{1-(1/G)^N}{1-1/G} \approx \frac{F}{1-1/G}.$$

Thus, although eqn (17.12) suggests that we can reduce F by reducing G, in fact, when we combine a number of stages to obtain a large over-all gain, the over-all noise figure is not improved. The noise measure $F/[1-(1/G)]$ is a better parameter than F. Because of this we shall assume that G is always large, so that (17.12) reduces to $F = 1+(T_m/T)$ and (17.13) to $T_n = T_m$.

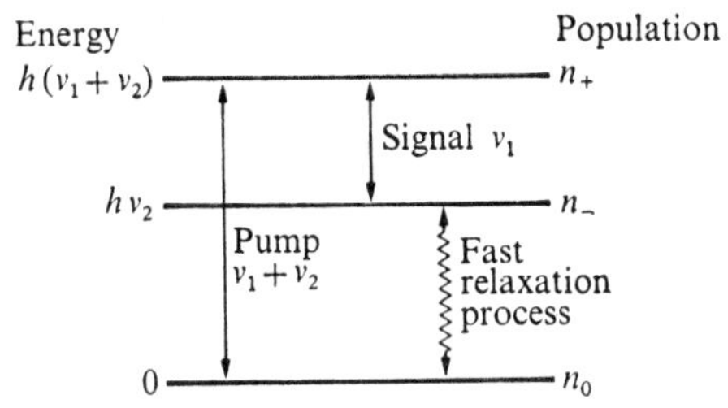

FIG. 17.2. Energy levels, populations, and transitions in a three-level maser.

With the same assumption, the basic expression (17.10) becomes

$$P_{si} = h\nu\,d\nu + \frac{h\nu\,d\nu}{\exp(h\nu/kT)-1} + \frac{h\nu\,d\nu}{\exp(h\nu/kT_m)-1}. \tag{17.14}$$

We see that even if $T = 0$ and $T_m = 0$ (which corresponds to a complete inversion of the maser level population with $n_- = 0$) the minimum detectable signal power is not $\frac{1}{2}h\nu\,d\nu$ but $h\nu\,d\nu$. This is truly detectable for there is no reason why the output of the maser should not be so large as to be a true, classical observable. There is little sense in defining a noise figure for this case for, although the input noise to signal ratio is apparently $\frac{1}{2}h\nu\,d\nu/P_{si}$, this is not observable. Thus, although it might appear from (17.4) that the limiting noise figure is $F = 2$, this is meaningless.

Practical masers usually operate some way away from this quantum limit, e.g. if the two temperatures involved in (17.14) are of the order of 4 K we have $h\nu \sim kT$ when $\nu \sim 10^{11}$ Hz, corresponding to a wavelength of 3 mm. It is therefore of some importance to understand the factors which influence the value of T_m. To discuss this we consider the three-level maser scheme shown in Fig. 17.2. A strong external pump signal at $\nu_1+\nu_2$ induces transitions between the outer levels at such a rate that it equalizes their populations n_+ and n_0. At the same time a fast thermal relaxation process maintains the relative populations of the two lower levels in the equilibrium ratio

$$n_-/n_0 = \exp(-h\nu_2/kT_0),$$

where T_0 is the ambient temperature. The ratio of the populations of the active levels is $n_+/n_- = n_0/n_- = \exp(h\nu_2/kT_0)$, and, since their energy separation is $h\nu_1$, the effective negative temperature characterizing these two levels is

$$T_m = T_0\nu_1/\nu_2. \tag{17.15}$$

In the low-frequency limit this is the noise temperature of the maser. We note that if $\nu_1 \ll \nu_2$ this can be much less than the ambient temperature of the device. There is no intrinsic reason why a device exchanging heat freely with surroundings at T_0 should not have a noise temperature below T_0. Another good example is provided by the f.e.t. At frequencies in the 1 MHz region its noise temperature may be several orders of magnitude less than the channel temperature. Unfortunately in a maser, to make $\nu_1 \ll \nu_2$ we have to make the pump frequency $\nu_1+\nu_2$ much greater than ν_1. If ν_1 is already a microwave frequency this may not be practicable.

If we neglect the thermal-noise power associated with the source, the signal power required for unity signal to noise ratio in a bandwidth $d\nu_1$ is

$$P_s = \left\{h\nu_1 + \frac{h\nu_1}{\exp(h\nu_1/kT_m)-1}\right\} d\nu_1 = \left\{h\nu_1 + \frac{h\nu_1}{\exp(h\nu_2/kT_0)-1}\right\} d\nu_1,$$

and even if $h\nu_1 \ll kT_0$ it is possible, if $\nu_2 \gg \nu_1$ for the zero-point term $h\nu_1\, d\nu_1$ to be significant.

Needless to say these considerations also apply to optical amplifiers using an active medium at optical (or infrared) frequencies, but so far the technology of these laser devices has not reached the point where noise performance is of practical significance.

It is interesting to compare the result (17.15) with the corresponding result (16.20) for a parametric amplifier. They are identical if we treat ν_2 as the idler frequency and T_0 as the temperature of the idler load. Indeed (Gordon Walker and Louisell 1963, Louisell 1964) the parametric amplifier can be shown to be closley related, in the principle of its operation to the maser.

Perhaps we may conclude our discussion of masers by remarking that our results apply only to the idealized core of the system. In practical masers, whether of the cavity or travelling-wave type, resonant circuits, feed lines, and circulators also may contribute thermal noise. Because the intrinsic noise of the maser is so low contributions of this type, which are negligible in other amplifiers, are of serious significance.

18 Oscillators

18.1. Introduction

OSCILLATORS may be classified in terms of whether their output is sinusoidal or non-sinusoidal. They may be classified also in terms of their internal structure. Thus we may distinguish between oscillators in which the frequency-selective element is a resonant circuit and oscillators in which it is a resistance–capacitance network; or we may distinguish between oscillators in which the amplitude is stabilized by a non-linear element which responds slowly to the mean amplitude over many cycles and oscillators in which the amplitude is stabilized by a fast-acting non-linear element which responds to the instantaneous amplitude within a single cycle. In this chapter we shall only consider sinusoidal oscillators containing a resonant circuit and a fast-acting non-linear element, since this corresponds to those oscillators in which noise is of practical significance.

Ideally, an oscillator produces a strong coherent sinusoidal output of constant amplitude, frequency, and phase. In practice, noise generated within the circuit superimposes a random background on this output. Thus, if the output $V(t)$ of an oscillator is observed over a long period T and we take as the unit of time $T/2\pi$, we can express $V(t)$ as

$$V(t) = \sum_{n=-\infty}^{\infty} V_n \exp(jnt). \tag{18.1}$$

In this series some pair of terms with $n = \pm n_0$ will correspond to the coherent output, and the other terms will represent noise. However, we could also express $V(t)$ as $V_0(t)\exp\{j\phi(t)\}\exp(jn_0 t)+$c.c. and regard fluctuations in $V(t)$ in terms of a random amplitude modulation $V_0(t)$ and phase modulation $\phi(t)$. Clearly, if the coefficients V_n in (18.1) are known they determine $V_0(t)$ and $\phi(t)$, and the statistical properties of $V_0(t)$ and $\phi(t)$ are related to those of the coefficients V_n. However, this relation involves the phases of the coefficients V_n as well as their amplitudes, and so knowledge of the power spectrum of the voltage $V(t)$ is not by itself enough to determine the properties of $V_0(t)$ and $\phi(t)$. Nevertheless, the intensity of the noise, i.e. its power spectrum, is sufficient to indicate the order of magnitude of the amplitude and phase modulation noise, and so we shall confine our attention to this aspect of the noise. For a fuller discussion the reader should consult Rowe (1965).

Any oscillator producing a stable output necessarily contains a frequency-selective circuit, a feedback loop with power gain or a device with a negative differential conductance, and a non-linear element to stabilize the amplitude. The presence of this non-linear element leads to serious mathematical problems in the exact analysis of either the steady coherent output or the noise, and indeed exact, or even reasonably accurate approximate, solutions can only be obtained for a few highly idealized systems, which may not correspond very accurately to real oscillators. Thus the comparison between the results of even the most sophisticated mathematical treatments and the behaviour of real oscillators leads to only qualitative agreement. Because of this we shall not attempt to give more than a simple heuristic, and qualitative, account of oscillator noise. This will suffice however to indicate the magnitude of the noise to be expected in particular cases and to suggest which circuit parameters are of most significance. The reader interested in a more formal mathematical treatment will find van der Pol's (1927) classic paper and the references associated with it in *Modern mathematical classics—analysis* (ed. R. Bellman 1961), a useful introduction.

The treatment of oscillators in electronic texts is generally rather brief and mainly concerned with questions such as power output, efficiency, and frequency stability, and so we shall preface our discussion of oscillator noise with a short account of oscillator behaviour in general.

18.2. Oscillator circuits

A few oscillators, e.g. tunnel-diode oscillators or the venerable dynatron oscillator using a tetrode, employ devices with a negative differential conductance as the active element, but the majority of practical oscillators employ an amplifying device and a positive feedback loop. A typical circuit, in this case due to Faulkner and Holman (1967), is shown in Fig. 18.1. We shall

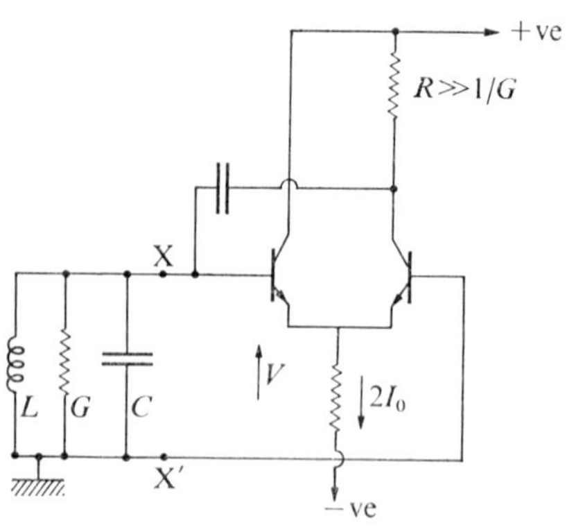

FIG. 18.1. Faulkner and Holman's oscillator.

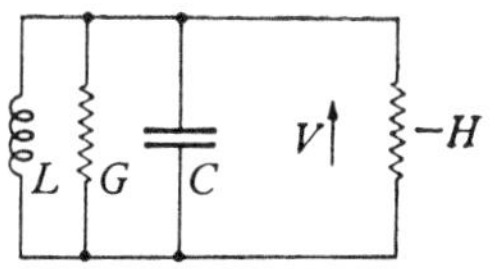

FIG. 18.2. The negative conductance oscillator.

return to this circuit at a later stage since, for varying values of the bias current I_0, it exemplifies almost all the significant features of oscillator behaviour. For our present purposes, however, it is enough to note that the active circuit makes contact with resonant circuit, *LC* with a loss conductance *G*, only across the terminals XX′. Therefore, provided that it is free from phase shift, it can be represented as in Fig. 18.2, where the negative conductance $-H$ represents the effect of the active network. Because *H* depends on the characteristics of the devices within the active network it will be non-linear, and we may express this by regarding it as a function $H(V)$ of the voltage.

An alternative, and occasionally more useful, but essentially equivalent representation is shown in Fig. 18.3. Here the output of the linear amplifier *A* drives a non-linear device *NL*, and the resulting output current *I* is fed back to the tank circuit to maintain oscillation. The element *NL* may be merely a convenient representation of an intrinsic non-linearity inherent in the amplifier *A* or it may actually represent a discrete device in the circuit. All practical amplifiers, including those with mutual inductance coupling, e.g. the tuned-anode triode oscillator, or split-capacitance coupling (e.g. the Colpitts oscillator), can be analysed in terms of either Fig. 18.2 or Fig. 18.3. Clearly, *H* in Fig. 18.3 is equal to I/V in Fig. 18.3. Both these figures correspond to oscillators in which the amplitude is regulated by a fast-acting non-linearity in the active element or network, but many practical oscillators also incorporate an additional slow-response stabilizing element which would, in Fig. 18.3, control the gain *A* of the amplifier in response to changes in the mean amplitude of oscillation. A typical example of this is the leaky-grid, or self-biased, triode oscillator. The primary fast control of amplitude is due to the

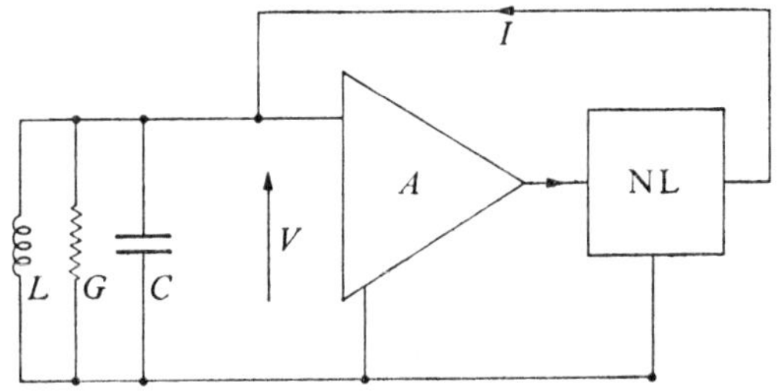

FIG. 18.3. A basic oscillator circuit.

non-linearity of the triode characteristic, and the slow control is due to the accumulating negative grid bias, which responds to the amplitude of a large number of cycles of oscillation.

18.3. Steady oscillations

If a parallel-tuned circuit LC, with loss described by a parallel conductance G, is set oscillating at its natural resonant frequency $\omega = (LC)^{-\frac{1}{2}}$ the amplitude of the oscillations will decay at a rate $\omega/2Q$, where $Q = \omega C/G$. If, however, a linear negative conductance $-H$ with $H > G$ is added in parallel to the circuit, the oscillations will increase in amplitude at a rate $\omega/2Q'$, where $Q' = \omega C/(H-G)$. Clearly, the oscillations will only settle down to a steady constant amplitude if H decreases with increasing voltage. Equally, if the amplifier network shown in Fig. 18.3 is connected to the tuned circuit, the oscillations will increase if $I > GV$ and will only be stabilized as a result of the presence of the non-linear element NL. If, for very small values of V, $H > G$ or $I > GV$, the oscillations will build up spontaneously from zero, or at least from an initial noise excitation. We shall always assume that this condition is fulfilled, although it should be noted that many oscillators, especially those constructed using bipolar transistors, may require a finite initial excitation to enter the regime where they sustain steady oscillations.

If steady sinusoidal, or approximately sinusoidal, oscillations with $V(t) = V_0 \sin \omega t$ are to exist, the average power $\frac{1}{2}GV_0^2$ dissipated in the loss G must be supplied by the active device or network. Thus, in terms of Fig. 18.2, we must have

$$\tfrac{1}{2}GV_0^2 = \frac{\omega}{2\pi}\int_0^{2\pi/\omega} H(V)V_0^2 \sin^2\omega t \, dt \tag{18.2}$$

or, in terms of Fig. 18.3,

$$\tfrac{1}{2}GV_0^2 = \frac{\omega}{2\pi}\int_0^{2\pi/\omega} I(V)V_0 \sin \omega t \, dt. \tag{18.3}$$

These equations determine the amplitude V_0 in terms of the behaviour of $H(V)$ or $I(V)$ with V. In general $H(V)$ and $I(V)$ will be complicated non-linear functions of V, but we may distinguish two extreme cases. In the first case, for any value of V that may occur in the circuit, $H(V)$ is a smooth continuous function of V and may be expanded in a rapidly-converging power series as

$$H = H_0 + H_1V + H_2V^2 + H_3V^3 + \text{etc.} \tag{18.4}$$

or alternatively

$$I = H_0V + H_1V^2 + H_2V^3 + \text{etc.} \tag{18.5}$$

An oscillator which can be so described is known as a van der Pol (1927) oscillator. The other extreme case occurs when $I(V)$ has a sharp knee at some

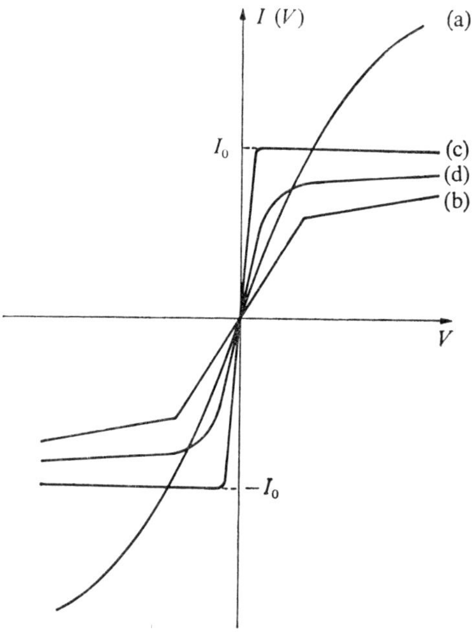

FIG. 18.4. Oscillator transfer characteristics (a) van der Pol; (b) Robinson; (c) extreme Robinson; (d) intermediate.

value of $|V|$. In the limit, $I(V) = I_0(V/|V|)$, where I_0 is a positive constant. An oscillator with a sharp knee in the characteristic is known as a Robinson (1959) oscillator. Fig. 18.4 shows $I(V)$ for (a) the van der Pol oscillator, (b) a Robinson oscillator, (c) the limiting case of a Robinson oscillator, and (d) an intermediate case. The oscillator shown in Fig. 18.1 has

$$I \sim I_0 \tanh\left(\frac{eV}{2kT}\right),$$

and for large V_0 approaches the behaviour of curve (c), while if $V \ll 2kT/e$, we have

$$I \sim I_0\left\{\frac{eV}{2kT} - \frac{1}{3}\left(\frac{eV}{2kT}\right)^3\right\}, \tag{18.6}$$

and the system is a van der Pol oscillator.

For a van der Pol oscillator eqn (18.2) yields

$$\tfrac{1}{2}GV_0^2 = \tfrac{1}{2}H_0V_0^2 + \tfrac{3}{8}H_2V_0^4 + \tfrac{5}{16}H_4V_0^6 + \text{etc.} \tag{18.7}$$

and, if this is to have a solution other than $V_0 = 0$, one at least of the even coefficients H_2, H_4, etc. must be negative, since H_0 must be positive and greater than G for oscillations to start spontaneously. Clearly, if H_2 is negative and succeeding coefficients are numerically small, oscillations will build up from zero until

$$V_0^2 = \frac{4}{3}\frac{H_0-G}{-H_2}. \tag{18.8}$$

For the extreme Robinson oscillator I (in Fig. 18.3) consists of a square wave of amplitude I_0 locked in frequency and phase to $V = V_0 \sin \omega t$, and eqn (18.3) yields

$$\tfrac{1}{2}GV_0^2 \equiv \frac{\omega}{\pi}\int_0^{\pi/\omega} I_0 V_0 \sin \omega t \, dt = \frac{2I_0 V_0}{\pi},$$

so that

$$V_0 = \frac{4I_0}{\pi G}. \tag{18.9}$$

As the bias current I in Fig. 18.1 is increased from zero, oscillations start when $I_0 = 2kTG/e$ and initially increase with I_0 as

$$V_0 = \frac{4kT}{e}\left(\frac{I_0-2kTG/e}{I_0}\right)^{\frac{1}{2}},$$

until I_0 exceeds about $4kTG/e$, when they approach (18.9) asymptotically.

The square-wave output of the feedback loop in the Robinson oscillator contains a high third-harmonic content of amplitude $(4/3\pi)I_0 = \frac{1}{3}GV_0$ but, if the Q of the resonant circuit is large, this produces little distortion of the voltage waveform. The amplitude of the third-harmonic voltage is

$$V^3 \sim V_0/9Q.$$

The non-linear term H_2 in the van der Pol oscillator also generates a third-harmonic component whose amplitude is

$$V_3 \sim \frac{V_0}{9Q}\frac{H_0-G}{G}$$

and so for marginal operation with $H_0-G \ll G$ the output of the van der Pol oscillator is the purer sine wave. On the other hand, if the term H_1V in the series expansion (18.4) is not zero, the van der Pol oscillator may produce an appreciable second-harmonic content in its output. In practice both circuits produce almost pure sine waves if $Q > 100$.

In practical Robinson oscillators the non-linearity is generally introduced as a specific feature of the design. Fig. 18.1 is a typical example, but similar behaviour can also be obtained with vacuum tubes (Robinson *loc. cit.*) or f.e.ts. In van der Pol oscillators, on the other hand, the non-linearity is usually an incidental property inherent in the active element or amplifier.

Now, in a van der Pol oscillator, the amplitude depends critically on H_0–G and H_2, and both H_0 and H_2 are in turn critically dependent on the device parameters. As a result, small changes in temperature, or bias, in the active device, produce disproportionately large changes in V_0. The behaviour of a van der Pol oscillator is therefore rather erratic and unpredictable, and practical van der Pol oscillators usually require an additional slow amplitude-stabilizing mechanism to yield reproducible results. This often leads to undesirable side-effects, such as intermittent oscillation (or squegging), if the subsidiary control mechanism is incorrectly designed. By contrast, Robinson oscillators have a predictable and reproducible performance. Because in the van der Pol oscillator small changes in H_0 produce large changes in output, flicker noise in the active device, which is to some extent equivalent to slow random changes in the transconductance of the device and is reflected in changes in H_0, produces related random changes in V_0. The van der Pol oscillator therefore tends to exhibit more amplitude and phase modulation noise than the Robinson oscillator. Apart from this the general behaviour of the two oscillators is in most other respects very similar, and most quantitative mathematical discussions of oscillator noise have been based on the properties of the van der Pol oscillator. We remind the reader again that few practical oscillators correspond to either of these extreme cases. We now discuss briefly why, although it is relatively easy to realize a practical van der Pol oscillator using vacuum tubes, it is much less easy to do so using transistors.

In most oscillators the magnitude of the negative conductance presented by the active network to the tank circuit is proportional to the transconductance g_m of the active devices in the network. If for one of these active devices $I = f(V, V_{dc})$, where V_{dc} is the bias voltage, then H_0 is proportional to dI/dV and H_2 to d^3I/dV^3. Thus, if the circuit is to behave as a van der Pol oscillator, dI/dV and d^3I/dV^3 must have opposite signs. In vacuum tubes obeying the Childs' three-halves power law, $I = K(V+A)^{\frac{3}{2}}$, where K and A are constants, and we have $dI/dV = K(V+A)^{\frac{1}{2}}$, while $d^3I/dV^3 = -\frac{3}{8}K(V+A)^{-\frac{3}{2}}$. Thus vacuum tubes, such as triodes and pentodes, are suitable for use in van der Pol oscillators. In f.e.t.s, on the other hand, $I = I_0[1+(V/V_p)]^2$ and d^3I/dV^3 and all higher derivatives vanish. Thus, if an f.e.t. oscillator produces a stable sinusoidal output it is either because additional non-linear elements are present or because the transistor displays small departures from the ideal characteristic $I = I_0[1+(V/V_p)]^2$. In the latter case the performance will be exceedingly unpredictable. For bipolar transistors $I = I_0 \exp(eV/kT)$ and all derivatives have the same sign. Thus an oscillator incorporating bipolar transistors is unlikely to run as a van der Pol oscillator unless the circuit is rather carefully designed as, for example, in Fig. 18.1. In general, if an oscillator, using transistors of either type, is not designed to ensure that it can operate as either a van der Pol or a Robinson oscillator, the amplitude will only be limited by clipping, when the voltage swing exceeds the available voltage supply. In this case it will not produce even approximately sinusoidal oscillations.

18.4. Forced oscillations

If either type of oscillator is driven from an external source at a frequency $\omega = \omega_0 + \delta\omega$, differing from the natural resonance frequency ω_0, the response will depend critically on the amplitude of the driving term and the frequency difference $\delta\omega$. The effect can be analysed rigorously for the van der Pol oscillator, and this is the substance of van der Pol's paper (*loc. cit.*), which has been the basis of all subsequent discussions of oscillator behaviour. However, we can also obtain a qualitative insight into the response as follows. In Fig. 18.5 the external source is represented by the current generator i. This current will stay in a sensibly constant phase relationship to the free oscillations at ω_0 over a period of time of the order of $1/\delta\omega$ and, during this

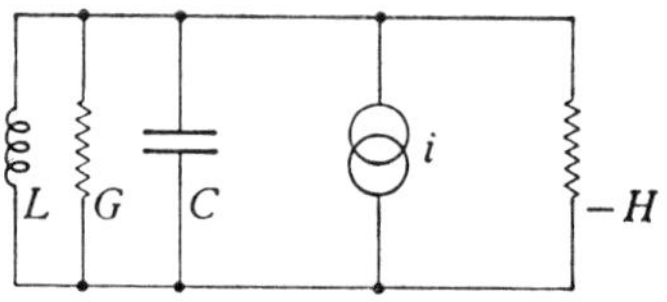

FIG. 18.5. Basic oscillator with a noise current source.

time, it will exchange an energy of the order of $\frac{1}{2}Vi_0/\delta\omega$ with the circuit, where V_0 and i_0 are the r.m.s. amplitudes of the free oscillation V and the perturbing current i. The factor $\frac{1}{2}$ arises because of the slowly-changing phase relation. If this energy is comparable to the energy CV_0^2 stored in the tuned circuit, the presence of i will have a profound effect on the nature of the oscillations and, as van der Pol shows, the free oscillations will be suppressed and the whole circuit will lock in phase and frequency to the driving term. Clearly, the frequency width of the locking region is given by

$$\delta f = \frac{\delta\omega}{2\pi} = \frac{i_0}{4\pi C V_0}. \tag{18.10}$$

This result, except for a numerical factor, is essentially independent of whether we are dealing with a van der Pol or a Robinson oscillator, or indeed an oscillator stabilized by a slow control mechanism.

The conductance G in Fig. 18.5 will generate a noise current of power spectrum $4kTG$ and, in addition, there will be noise due to the active device represented by the negative conductance H. We express the total noise power spectrum of i as $4nkTG$, where $n > 1$.

The influence of a coherent signal on the free oscillations is only appreciable when the energy it contributes to the circuit, before phase coherence is lost, is comparable with the energy of the free oscillations. The total energy associated with the noise is about nkT, and so the size of this energy compared with CV_0^2 clearly will be important. The ratio nkT/CV_0^2, which is the ratio of the mean square noise voltage to V_0^2, also appears in another context if we assume that the noise current is only significant in a band of width f_0/Q, and insert its r.m.s. value for this band in (18.10) to obtain the approximate width of the locking region as

$$\delta f = \frac{(4nkTGf_0/Q)^{\frac{1}{2}}}{4\pi CV_0}. \tag{18.11}$$

If δf, from this calculation, is less than f_0/Q the noise can hardly 'capture' the free oscillation. Thus, if

$$\frac{4nkT\omega}{(4\pi CV_0)^2} < \frac{f_0}{Q}$$

or

$$\frac{nkT}{2\pi CV_0^2} < 1, \tag{18.12}$$

the noise is unlikely to make any serious difference to the character of the free oscillation. Now with typical values $n \sim 10$, $T \sim 300$ K, and $C \sim 10^{-10}$ F this condition is

$$\frac{10^{-10}}{V_0^2} < 1,$$

and so, for any reasonable amplitude V_0, we can assume that the noise has very little effect on the free oscillation. The total output of the circuit will consist of a more or less unchanged free oscillation, together with a more or less independent random background. We can calculate this background by considering the response of the system to individual noise components without taking the possibility of strong interaction, or locking, into account.

18.5. The noise output

If the parallel resonant circuit alone is driven by the noise current the power spectrum of the voltage developed across the circuit is

$$W = \frac{4nkTG}{G^2+(2\pi f_0 C)^2(f/f_0-f_0/f)^2}, \tag{18.13}$$

and, provided we ignore the possibility of locking, this is also valid if G in the denominator corresponds to a negative conductance. Thus, if H were linear and equal to H_0 we should obtain

$$W = \frac{4nkTG}{(H_0-G)^2+(2\pi f_0 C)^2(f/f_0-f_0/f)^2}.$$

In a van der Pol oscillator the negative conductance H is not, however linear, and the response of H due to a voltage at f is modified by the presence of the large free-oscillation amplitude V_0. The effect is complex, but when

$$|f-f_0| \gg f_0/Q$$

it is small and when $|f-f_0| \ll f_0/Q$ we have to replace (H_0-G) by $2(H_0-G)$. The influence of V_0, at $\omega_0 = 2\pi f_0$, on a component v at ω arises essentially because $H_2V^3 = H_2(V_0 \sin \omega_0 t + v \sin \omega t)^3$ contains a term

$$V_0^2 v \sin^2 \omega_0 t \times \sin \omega t = \tfrac{1}{2}V_0^2 v(1-\cos 2\omega_0 t) \sin \omega t,$$

and this contains a component at ω.

The situation in the Robinson oscillator is somewhat simpler. A small voltage $v \sin \omega t$ across the tank coil produces only a small phase modulation of the fed-back current pulses, and this modulation is linear in v. The resultant change in the voltage developed across the tank circuit is quadratic in the phase modulation and is negligible. Thus in this case the noise voltage power spectrum is given simply by the direct effect of i on the tank circuit and therefore corresponds to (18.13). We are interested primarily in the noise near resonance, where $|f-f_0| \ll f_0/Q$ and, for the van der Pol oscillator, this is

$$W_p = \frac{4nkTG}{4[H_0-G)^2+(2\pi f_0 c)^2(f/f_0-f_0/f)^2} \approx \frac{4nkT}{G}\left(\frac{G}{2(H_0-G)}\right)^2. \quad (18.14)$$

For the Robinson oscillator

$$W_r = 4nkT/G. \quad (18.15)$$

Eqn (18.14) can also be written as

$$W_p \approx \frac{4nkT}{G}\left(\frac{2G}{3H_2V_0^2}\right)^2, \quad (18.16)$$

and this indicates that the noise output decreases as V_0 increases. This behaviour is observed experimentally, but the initial decrease of the noise, from very high values at low values of V_0, does not persist indefinitely as V_0 is increased. The reason is relatively simple. If $-H_2V_0^2$ exceeds G, the effect of higher-order terms in the expansion of H as a power series in V cannot be neglected. To neglect them is only valid when $-H_2V_0^2 = \tfrac{4}{3}(H_0-G)$ is small compared with H_0 and G (because $H_0 \sim G$). Thus, as V_0 is increased until $-H_2V_0^2 \sim G$ (by increasing H_0 or decreasing G), the behaviour of the van der Pol oscillator begins to approximate to that of the Robinson oscillator and the noise never falls below the value given by (18.15). The ratio of the noise to the coherent output is, however, proportional to W/V_0^2 and decreases as V_0 increases. Thus, when an oscillator with a particularly pure sinusoidal output is required, it is advantageous to drive it hard, i.e. well beyond the van der Pol region, to increase the Q of the tank circuit, which decreases G, and to ensure that the active network does not lead to a high value of n. With this in mind we now consider the factors that determine the noise coefficient n.

18.6. The noise coefficient *n*

This coefficient, or rather $(n-1)4kTG$, represents noise added by the active device and its associated network of components. If the active device is a true negative-conductance element the value of n depends on physical processes within the device. We shall consider one such example, the reflex klystron, at a later stage. To begin with, we consider oscillators in which the power gain is provided by an amplifier and a feedback loop, and we shall have to consider van der Pol and Robinson oscillators separately. Fig. 18.6

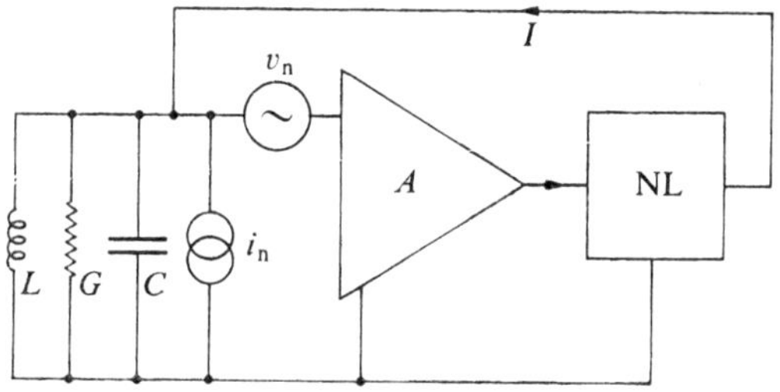

FIG. 18.6. Noise sources v_n and i_n associated with an oscillator circuit.

shows the basic configuration together with the two noise generators i_n and v_n associated with the amplifier. Clearly i_n is simply an addition to the Johnson noise due to G, but v_n needs separate consideration. Its effects appear in the feedback current I as Hv_n. Thus the total power spectrum of the current acting on the tank circuit is $4kTG+w_i+H^2w_v$ and, if we express w_i as $4kTg_n$ and w_v as $4kTr_n$, we have

$$n = 1+g_n/G+Gr_n(H/G)^2. \tag{18.17}$$

Unless the oscillator is driven hard, $H \sim H_0 \sim G$, and we have

$$n \approx 1+g_n/G+Gr_n, \tag{18.18}$$

which is just the noise figure F of the amplifier A with a source of internal impedance $1/G$. For example, if the first stage of the amplifier is a bipolar transistor with a collector current I_c and $g_m = eI_c/kT$ we have $g_n = \frac{1}{2}g_m\beta^{-1}$ and $r_n = \frac{1}{2}g_m^{-1}$, where β is the d.c. current gain. Suppose that $\beta = 100$,

$I_c = 1$ mA, and $g_m = 40$ mA/v then, if $1/G = 10^4\ \Omega$, we have

$$n = 1+2+10^{-3} \sim 3.$$

If, however, Fig. 18.6 represents a Robinson oscillator, the effects of i_n are still additive, but v_n no longer produces amplitude modulation of the feedback current I. Its effects are apparent as phase modulation. Consider the effect of a component $v \sin\{(\omega_0+\delta\omega)t+\phi\}$ added to the free oscillation $V_0 \sin \omega_0 t$. The voltage at the amplifier input is

$$\begin{aligned} V &= V_0 \sin \omega_0 t + v \sin\{(\omega_0+\delta\omega)t+\phi\} \\ &= \{V_0^2+2V_0 v \cos(\delta\omega t+\phi)+v^2\}^{\frac{1}{2}} \sin\left\{\omega_0 t + \tan^{-1}\frac{v \sin \delta\omega t}{V_0+v \cos \delta\omega t}\right\}, \end{aligned}$$

which can be approximated as

$$V \sim \{V_0^2+2V_0 v \cos(\delta\omega t+\phi)+v^2\}^{\frac{1}{2}} \sin\left\{\omega_0 t+\frac{v}{V_0} \sin \delta\omega t\right\}.$$

Provided that $\delta\omega \ll \omega_0$ the current fed back to the input circuit has a slow phase modulation and, if the square-wave amplitude is I_0, the fundamental component near ω_0 is

$$I = \frac{4I_0}{\pi} \sin\left(\omega_0 t+\frac{v}{V_0} \sin \delta\omega t\right) \approx \frac{4I_0}{\pi}\left(\sin \omega_0 t+\frac{v}{V_0} \cos \omega_0 t \sin \delta\omega t\right).$$

Since $V_0 = \left(\frac{4}{\pi}\right)(I_0/G)$ the additional term in I is

$$i' = vG \cos \omega_0 t \sin \delta\omega t = \tfrac{1}{2} vG\{\sin(\omega_0+\delta\omega)t - \sin(\omega_0-\delta\omega)t\}.$$

A Fourier component of the amplifier noise at $\omega_0+\delta\omega$ produces correlated noise sidebands at $\omega_0+\delta\omega$ and $\omega_0-\delta\omega$. Clearly, the simplest description of the effect of amplifier noise is as phase modulation of V_0, and we shall discuss this aspect shortly, but for the moment we forget the correlation and concentrate our attention on the power spectrum of i'. Noise at $\omega_0+\delta\omega$ arises from noise in v_n at $\omega_0 \pm \delta\omega$, and so

$$w_i' = 2G^2(w_v/4) = \tfrac{1}{2}G^2 w_v = \tfrac{1}{2}G^2 4kTr_n.$$

Thus

$$n = 1+g_n/G+\tfrac{1}{2}Gr_n. \tag{18.19}$$

Apart from the factor $\frac{1}{2}$ in the last term this is identical with the result for the van der Pol oscillator, and practical oscillators are probably quite adequately described by either formula.

The amplitude of the phase modulation of V_0 due to the component of amplitude v is v/V_0 near resonance but, if the Q of the tank circuit is high, then even when $\delta\omega \ll \omega_0$ the impedance of the tank circuit is not $1/G$ but

$$Z = \frac{1/G}{1+jQ(\omega/\omega_0-\omega_0/\omega)}$$

and the phase modulation is of amplitude

$$\phi(v) = \frac{v/V_0}{\{1+Q^2(\omega/\omega_0-\omega_0/\omega)^2\}^{\frac{1}{2}}}.$$

We now introduce the power spectrum $4kTr_n$ of v and obtain the mean square phase modulation due to all the frequency components of v as

$$\langle\phi^2\rangle = \frac{1}{V_0^2}\int_0^\infty \frac{4kTr_n\,df}{1+Q^2(f/f_0-f_0/f)^2} = \frac{\pi}{2}\frac{4kTr_nf_0}{QV_0^2},$$

which can be expressed as

$$\langle\phi^2\rangle = \frac{kT}{CV_0^2}Gr_n.$$

If, for example, $T = 300$ K, $G^{-1} = 10^4\,\Omega$, $r_n = 100\,\Omega$, and $C = 100$ pF, we have

$$\langle\phi^2\rangle \sim 4\times10^{-13}/V_0^2.$$

For $V_0 = 1$ V the r.m.s. phase deviation is about 7×10^{-7} rad, and this also corresponds roughly to fractional frequency deviation $\Delta f/f_0$. This calculation only takes the noise due to voltage noise in the amplifier into consideration. To it we must add the randomness in the phase due to current noise in the tank circuit and this, in either the van der Pol or Robinson oscillator, brings the total phase uncertainty to

$$\Delta\phi^2 \sim \frac{nkT}{CV_0^2}. \tag{18.20}$$

In some practical oscillator circuits, e.g. the Colpitts circuit shown in Fig. 18.7, only a fraction α, in this case $\alpha = C_2/(C_1+C_2)$, of the voltage across the tank circuit is applied to the active device. This is, however, equivalent to

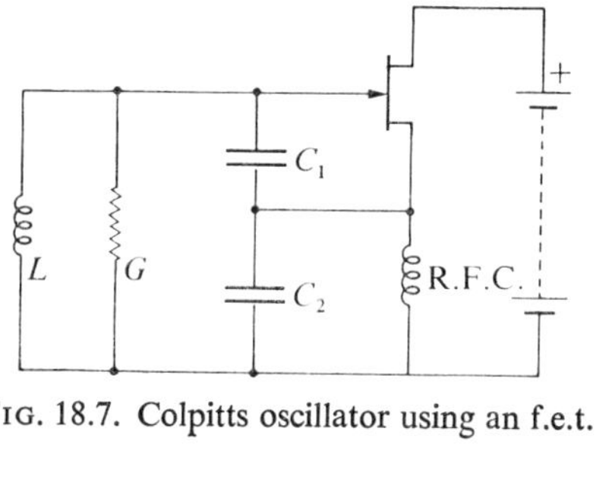

FIG. 18.7. Colpitts oscillator using an f.e.t.

an impedance transformation $G \rightarrow G/\alpha^2$ in terms of Fig. 18.6, and so the results that we have obtained will be valid if we simply replace G by this new value. The result may be a decrease or an increase in n, depending on whether α^2/G is nearer, of further, than $1/G$ from the value of the optimum source resistance for the active device.

In general, except in nuclear resonance spectrometers and where an oscillator is required as a local oscillator in a super-heterodyne system, noise in oscillators is not of much practical consequence. We do see, however, that, as a general rule, whenever a few millivolts of signal are required, it is better to obtain it by attenuating a larger signal from a more powerful oscillator than to attempt to design a low-level oscillator.

18.7. Reflex klystrons

Many microwave receivers are simple superheterodyne systems without r.f. amplification, and the first stage consists of a crystal mixer-diode driven by a reflex-klystron local oscillator. If f_0 is the local-oscillator frequency and f_i the intermediate frequency (i.f.), klystron noise at $f_0 \pm f_i$ will be accepted by the i.f. amplifier.

The main source of noise in a reflex klystron is shot noise in the electron beam. The effective beam current I_0 is usually somewhat less than the cathode current, since not all the emitted electrons traverse the resonant cavity, furthermore, the beam is partially space-charge smoothed so that the power spectrum of the beam-current fluctuations may be somewhat less than $2eI_0$. If the loaded-cavity quality factor is Q and the effective shunt impedance at resonance is Z, the power spectrum of the output noise, except very near f_0, is

$$W_n \sim 2eI_0 Z\{1+Q^2(f/f_0 - f_0/f)^2\}^{-1}, \tag{18.21}$$

and the noise energy asociated with the cavity is $2eI_0Z$. This is to be compared with the coherent energy $QP/2\pi f$ associated with an output power P. The ratio of these two terms determines the influence exerted by the noise on the behaviour of the oscillator. In a typical case with $I_0 = 15$ mA, $Z = 10^4\ \Omega$, $Q = 1000$, $f_0 = 10^{10}$ Hz, and $P_0 = 30$ mW the ratio is about 10^{-7}, and so the noise has little effect on the coherent oscillation. If, however, this oscillator is used with an intermediate frequency of 50 MHz the sum of the power spectra near $f_0 + f_i$ is $4eI_0Z/100 \sim 250\,kT$, which is substantial. Even though only about 1 mW of local oscillator power need be coupled to the diode, the attenuated noise, of power spectrum $10\,kT$, reaching the diode is still significant. In receivers where a low noise figure is required it is almost essential to use some sort of balanced mixer circuit to eliminate or minimize local oscillator noise. We note also that the use of as high an intermediate frequency as possible is also advantageous.

19 Mixers and phase-sensitive detectors

19.1. Introduction

A MIXER converts an input signal, with Fourier components centred on a carrier frequency f_r, to a proportionate signal centred on a different, usually lower, intermediate frequency f_i. This is achieved by beating the input signal with a local oscillator of frequency f_0, which may be either the sum frequency f_r+f_i or the difference frequency f_r-f_i. If the mixer follows an amplifier of substantial gain, noise generated in the mixer and the conversion gain or loss of the mixer are not significant. However, if, as in many microwave receivers, the mixer is the first stage, both these factors and, in addition, noise in the local oscillator drive have to be considered.

A phase-sensitive detector is essentially a mixer in which the local oscillator frequency f_0 coincides with the carrier frequency f_r, and the output is confined to very low frequencies near d.c. The magnitude and sign of the d.c. output depend on the relative phases of the carrier and the local oscillator, or reference, drive. The primary use of the phase-sensitive detector is in systems where we know *a priori* that any significant incoming signal will have a definite phase relation to an available reference source. The phase-sensitive detector then provides a means of identifying this signal, even if it is submerged in a much larger background of noise at other frequencies.

19.2. The mixing process

A mixer may be considered as a system whose output $I(t)$ can be expressed as the product

$$I(t) = L(t)V(t). \tag{19.1}$$

of the incoming signal $V(t)$ with a local oscillator drive function $L(t)$, of a definite fundamental period f_0. It is not necessary for $L(t)$ to be sinusoidal, in many cases it is a square wave, and in general we shall have

$$L(t) = L_0+\sum_{n=1}^{\infty}\{L_n \exp(2\pi jnf_0t)+\text{c.c.}\}. \tag{19.2}$$

Fig. 19.1 shows a typical mixer configuration in which the output of the mixer is fed to a band-pass filter, or amplifier, which accepts only Fourier components in a band Δf_i centred on f_i. The frequencies f_0, f_i, and Δf_i are

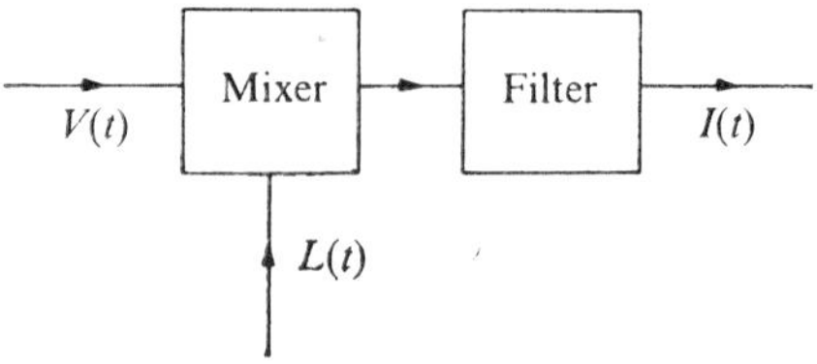

FIG. 19.1. Basic mixer configuration.

usually chosen so that $\Delta f_i \ll f_i$ and $f_i < f_0/2$. Thus, if we express the input voltage as

$$V(t) = \int_0^\infty \{V(f)\exp(2\pi jft) + \text{c.c.}\}\, df, \tag{19.3}$$

the output $I(t)$ contains only terms which arise as difference beats between components of $V(t)$ and the fundamental component of $L(t)$ with $n = 1$. Thus if

$$I(t) = \int_0^\infty \{I(F)\exp(2\pi jFt) + \text{c.c.}\}\, dF, \tag{19.4}$$

we have

$$I(F) = L_1 V^*(f_0 - F) + L_1^* V(f_0 + F). \tag{19.5}$$

It follows immediately that if $w(f)$ is the power spectrum of $V(t)$ and $W(F)$ is that of $I(t)$ then

$$W(F) = L_1 L_1^*\{w(f_0 - F) + w(f_0 + F)\}. \tag{19.6}$$

The signal content of $V(t)$ will be confined to a region of width f_i about either $f_0 - f_i$ or $f_0 + f_i$, for definiteness we assume that the signal channel is at $f_0 - f_i$, so that $f_0 + f_i$ is the image channel. We then see that the mixer output contains noise from both the signal and image channels. If the input $V(t)$ is unfiltered, so that w is white, the output noise to signal ratio is doubled by the image-channel noise. It is therefore desirable, and usually practicable, to restrict the bandwidth of the input amplifier so that its output does not contain amplified image channel noise. When this is done Δf_i determines both the signalling bandwidth and the noise bandwidth.

19.3. Phase-sensitive detection

In a phase-sensitive detector the pass-band of the input circuit generally extends down to frequencies approaching zero or at any rate well below $f_r = f_0$, and the output is fed to a low-pass filter extending down to d.c. The low-frequency output is given by

$$I(F) = L_0 V(F) + \sum_{n=1}^{\infty} \{L_n V^*(nf_0 - F) + L_n^* V(nf_0 + F)\}, \tag{19.7}$$

and the signal or d.c. term is

$$I_0 = L_1 V^*(f_0) + L_1^* V(f_0). \tag{19.8}$$

Practical phase-sensitive detectors are designed so that L_0, and all terms with n even, are zero; but if the input contains noise near $3f_0$, $5f_0$, etc. this noise, in addition to the noise near the signal at f_0, will appear in the output. Thus it is usual to filter the input so that $w(f)$ falls rapidly to zero when $f > f_0$. If $w(f)$ is white in the vicinity of f_0, the output noise power spectrum for low F is

$$W(F) = L_1 L_1^* \{w(f_0 - F) + w(f_0 + F)\} = 2L_1 L_1^* w. \tag{19.9}$$

We may take L_1 to be real and express the input signal in terms of its r.m.s. value V_s as

$$V_s(t) = \frac{V_s}{\sqrt{2}}\{\exp \mathrm{j}(2\pi f_0 t + \phi) + \text{c.c.}\} \tag{19.10}$$

and then the d.c. output signal is

$$I_0 = 2^{\frac{1}{2}} V_s L_1 \cos \phi. \tag{19.11}$$

When $\phi = 0$ the mean square output noise to signal ratio is

$$\left(\frac{N}{S}\right)^2 = \frac{1}{2V_s^2} \int^{\Delta F} 2w \, \mathrm{d}f = \frac{w \, \Delta F}{V_s^2}, \tag{19.12}$$

and so the noise bandwidth of the system is the noise bandwidth of the output filter. If, for example, this is a simple RC network of time constant τ we have $F = \frac{1}{4}\tau$.

In a typical application we might have $\tau = 100$ s while the input bandwidth might be 100 Hz, thus, if the r.m.s. signal to noise ratio after detection were unity, the ratio before detection would be $\frac{1}{200}$. If the input r.m.s. noise exceeds the signal by a factor of 200 the peak noise voltage will exceed it by perhaps 1000. This leads to an important practical problem in phase-sensitive detector design. If the detector is to work with an extreme ratio of the input and output bandwidths, it must not overload with noise peaks greatly exceeding the expected signal. If, in the absence of noise, the maximum output signal without overloading is $I_{\max}$ and, due to imperfections in design, the d.c. output drift is δI, the weakest signal detectable is related to the overload input by $\delta I / I_{\max}$. This ratio also determines the smallest useful value of $(4\tau\Delta f)^{-\frac{1}{2}}$.

19.4. Mixer input stages

Up to about 1000 MHz low-noise amplification is relatively straightforward and inexpensive, and it is unusual to find a mixer constituting the input stage of a system. At microwave frequencies, however, low-noise amplification is more difficult and, except in the most stringent applications, receivers tend to consist of a diode mixer followed by i.f. amplification. The intermediate frequency is usually a small fraction of the local-oscillator frequency and, since the low-power local oscillators used in microwave receivers are noisy, oscillator noise at both the signal frequency f_0+f_i and the image frequency f_0-f_i is significant. However, it can be reduced greatly by using two mixer diodes in a balanced configuration. If the electrical paths from the local oscillator to the mixers are not too different, their i.f. noise outputs due to oscillator noise will be correlated and in phase. If, however, an additional phase shift π is introduced in the signal path to one mixer, their signal outputs will be in antiphase. Thus, if the mixer outputs are connected in opposition to the input of the i.f. amplifier, their signal outputs will add and local-oscillator noise will be eliminated. This technique is almost universally employed in low-noise microwave receivers and reduces local-oscillator noise to tolerable proportions.

A diode mixer has a conversion gain G, defined as the ratio of i.f. output power to r.f. input power, which is less than unity. In addition, the mixer itself will generate noise and, if we describe this in terms of a power spectrum kT_d at the intermediate frequency, the over-all noise temperature of a receiver will be

$$T_n = T_s + \frac{1}{G}(T_d + T_a), \tag{19.13}$$

where T_a is the i.f. amplifier noise temperature and the additional source temperature term T_s comes from the converted image-channel noise. The values of the parameters G and T_d depend on a number of factors, including the nature of the diode. The gain G for a given diode depends not only on the impedances presented to the diode at the signal, image, and local-oscillator frequencies but also at harmonics of the local-oscillator frequency. It also depends on the variable capacitance of the diode which confers on the system some of the properties of a parametric amplifier. Under some circumstances G can exceed unity, but with present diodes this is not usually associated with low values of T_d. The value of T_d depends to some extent on the local-oscillator power and, since flicker or $1/f$ noise is significant, also on the intermediate frequency. Until recently, values of G below $\frac{1}{4}$ and of T_d above 300 K were common, but the Schottky-barrier diodes now available lead to values of G which are appreciably larger, and the best mixers have noise temperatures (excluding the image term) of the order of 300 K. Dragone (1972) has given a recent review. It should be noted, however, that even 300 K is considerably greater than the noise temperatures available with maser, or even parametric, amplifiers and that in the most critical applications, e.g. to satellite communication, these devices are to be preferred.

20 Detectors

MOST practical detectors fall into one of two categories: square-law detectors or linear detectors characterized by the law relating their output current to their input voltage

$$I = \alpha V^2, \tag{20.1S}$$

or

$$\left.\begin{aligned} I &= \alpha V & V &\geqslant 0 \\ I &= 0 & V &\leqslant 0 \end{aligned}\right\}. \tag{20.1L}$$

All detectors at low level behave as square-law detectors, bolometers at all levels are square-law detectors, crystals and diodes at high level approximate the behaviour of an ideal linear detector.

For detectors following high-gain amplifiers only the effect of the detector on the noise that already exists is required, but for detectors constituting the first stage of a receiver the noise generated in the detector is also important.

20.1. The detection process

A typical system involving a detector is shown in Fig. 20.1. The input consists of $V(t)$ containing noise and signals in a band Δf_0 centred on f_0, and the output of interest is at low frequencies from d.c. to Δf_0. The bandwidth Δf_0 need not be well defined, all we require is that the input is negligible at frequencies outside the limits, say $f_0 \pm 2\,\Delta f_0$, and that Δf_0 is less than about $f_0/4$.

The low-frequency output of the detector then contains components from d.c. to Δf_0 arising from difference beats between input components. The output also contains sum beats near f_0, $2f_0$, etc., which we shall neglect.

The properties of the low-frequency output are discussed most conveniently in terms of the envelope of the input. Let the input be expanded as a Fourier series in an interval T so that, adopting a unit of time $T/2\pi$, we have

$$V(t) = \sum_{n=0}^{\infty} a_n \cos(nt + \phi_n), \tag{20.2}$$

then only those coefficients a_n for which

$$f_0 - \frac{\Delta f_0}{2} < \frac{n}{T} < f_0 + \frac{\Delta f_0}{2}$$

have values appreciably different from zero.

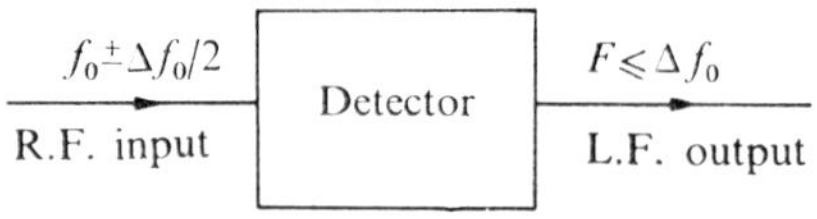

FIG. 20.1. Basic detector configuration.

By a little trigonometric manipulation we can express $V(t)$ as

$$V(t) = A \cos n_0 t - B \sin n_0 t, \tag{20.3}$$

where

$$A = \sum a_n \cos\{(n-n_0)t+\phi_n\}, \tag{20.3a}$$

$$B = \sum a_n \sin\{(n-n_0)t+\phi_n\}, \tag{20.3b}$$

and

$$n_0 = f_0 T \tag{20.3c}$$

corresponds to the centre frequency.

We can then write $V(t)$ as

$$V(t) = R(t)\cos(n_0 t+\theta), \tag{20.4}$$

where

$$R(t) = +\sqrt{(A^2+B^2)} \tag{20.4a}$$

and

$$\tan\theta = A/B. \tag{20.4b}$$

Since n ranges only over the limited interval $(f_0 \pm \Delta f_0/2)T$ no component of A or B varies more rapidly with t than $\Delta f_0/2$ and no component of $R(t)$ more rapidly than Δf_0. Thus $R(t)$ is a slowly varying function of t compared with $\cos n_0 t$. It is the envelope of $V(t)$ shown in Fig. 20.2.

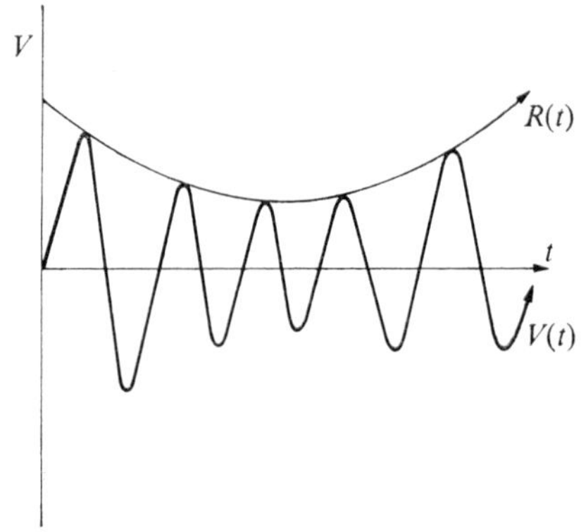

FIG. 20.2. The envelope $R(t)$ of an amplitude modulated signal $V(t)$.

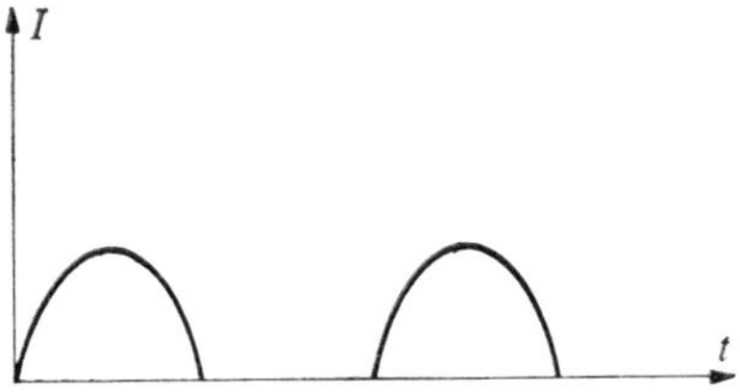

FIG. 20.3. Sine wave after linear detection.

A signal $V_0 \cos n_0 t$ applied to a square-law detector results in an output current

$$I = \alpha V_0^2 \cos^2 n_0 t = \tfrac{1}{2}\alpha V_0^2(1-\cos 2n_0 t), \tag{20.5S}$$

and the sole low-frequency component is the d.c. term

$$I_{\mathrm{dc}} = \frac{\alpha}{2} V_0^2. \tag{20.6S}$$

If the signal is $V(t) = R(t)\cos(n_0 t+\theta)$ the low-frequency output is clearly

$$I_{\mathrm{lf}} = \frac{\alpha}{2} R^2(t). \tag{20.7S}$$

A signal $V_0 \cos n_0 t$ applied to a linear detector results in the output shown in Fig. 20.3, whose sole low-frequency component is again a d.c. term

$$I_{\mathrm{dc}} = \frac{\alpha}{\pi} V_0, \tag{20.6L}$$

and so the low-frequency output for an input $R(t)\cos(n_0 t+\theta)$ is

$$I_{\mathrm{lf}} = \frac{\alpha}{\pi} R(t). \tag{20.7L}$$

20.2. Noise in a square-law detector

The output of a square-law detector is given by eqn (20.7S), which we may write as

$$I_{\mathrm{lf}} = \frac{\alpha}{2}(A^2+B^2) = \frac{\alpha}{2}\sum_n \sum_m a_n a_m \cos\{(n-m)t+\phi_n-\phi_m\}. \tag{20.8S}$$

If we expand the output as a Fourier series

$$I_{\mathrm{lf}} = \sum_k C_k \cos(kt+\theta_k), \tag{20.9}$$

then terms of frequency k arise from $m = n \pm k$; in particular, the d.c. term is

$$I_{\mathrm{dc}} = C_0 = \frac{\alpha}{2}\sum_n a_n^2. \tag{20.10S}$$

If the input is a noise process with a power spectrum $w(f)$ we have

$$\sum \langle a_n^2 \rangle = 2 \int_{f_0-\Delta f_0/2}^{f_0+\Delta f_0/2} w(f)\,df,$$

and so

$$I_{dc} = C_0 = \alpha \int_{f_0-\Delta f_0/2}^{f_0+\Delta f_0/2} w(f)\,df. \tag{20.11S}$$

The term of frequency k is given by

$$C_k \cos(kt+\theta_k)$$
$$= \frac{\alpha}{2}\left\{\sum_n a_n a_{n-k} \cos(kt+\phi_n-\phi_{n-k}) + \sum_n a_n a_{n+k} \cos(kt+\phi_{n+k}-\phi_n)\right\}.$$

The two sums in this expression yield identical results, for they differ only for terms with $n < k$, i.e. $f < \Delta f_0$, where the input is in any case zero. We therefore have

$$C_k \cos(kt+\theta_k) = \alpha \sum_n a_n a_{n+k} \cos(kt+\phi_{n+k}-\phi_n). \tag{20.12S}$$

The phases ϕ_{n+k} and ϕ_n are independently and randomly distributed, and so the ensemble average of this expression is zero. Thus

$$\langle C_k \rangle = 0, \qquad k \neq 0.$$

If we square eqn (20.12S) we have

$$C_k^2 \cos^2(kt+\theta_k)$$
$$= \alpha^2 \sum_n \sum_m a_n a_m a_{n+k} a_{m+k} \cos(kt+\phi_{n+k}-\phi_n)\cos(kt+\phi_{m+k}-\phi_m).$$

In an ensemble average the cross-terms with $m \neq n$ drop out, again because of the independence of the phases, and we are left with

$$\langle C_k^2 \rangle = \alpha^2 \sum_n \langle a_n^2 a_{n+k}^2 \rangle.$$

Since the a_n with different n are independent this is equivalent to

$$\langle C_k^2 \rangle = \alpha^2 \sum_n \langle a_n^2 \rangle \langle a_{n+k}^2 \rangle.$$

We now express $\langle C_k^2 \rangle$ in terms of a power spectrum $W(F)$, where $F = k/T$, and we then have

$$\langle C_k^2 \rangle = \lim_{T\to\infty} 2W(F)/T,$$

$$\langle a_n^2 \rangle = \lim_{T\to\infty} 2w(f)/T.$$

Sums over n become T times integrals over f, i.e.

$$\sum_n \rightarrow T \int \mathrm{d}f,$$

so that

$$W(F) = 2\alpha^2 \int_{f_0-\Delta f_0/2}^{f_0+\Delta f_0 f2} w(f)w(f+F)\,\mathrm{d}f. \tag{20.13S}$$

We have thus succeeded in expressing the output power spectrum in terms of the input spectrum.

20.3. Noise in a linear detector

No brief elementary treatment of this problem can be given because of the square root in the definition of $R(t)$. An exact treatment is given by Rice (1944), whose results we quote. The d.c. output is given by

$$I_{\mathrm{do}} = \alpha\left\{\frac{w(f)}{2\pi}\,\mathrm{d}f\right\}^{\frac{1}{2}} \tag{20.11L}$$

and the power spectrum by

$$W(F) = \frac{\alpha^2 \int w(f)w(f+F)\,\mathrm{d}f}{4\pi \int w(f)\,\mathrm{d}f}. \tag{20.13L}$$

20.4. White noise input in Δf_0

When the input consists of white noise of constant power spectrum w in Δf_0 these results lead to

square law,

$$I_{\mathrm{dc}} = \alpha w\,\Delta f_0 \tag{20.14S}$$

$$W(F) = 2\alpha^2 w^2\,\Delta f_0\left(1-\frac{F}{\Delta f_0}\right) \tag{20.15S}$$

linear

$$I_{\mathrm{dc}} = \alpha\left(\frac{w\,\Delta f_0}{2\pi}\right)^{\frac{1}{2}} \tag{20.14L}$$

$$W(F) = \frac{\alpha^2 w}{4\pi}\left(1-\frac{F}{\Delta f_0}\right). \tag{20.15L}$$

Thus for both detectors the noise spectrum has its maximum intensity near d.c. and falls to zero at $F = \Delta f_0$.

The total low-frequency noise output is obtained from eqn (20.15S,L) by integrating from d.c. to $F \geqslant \Delta f_0$ and is in each case

square law

$$\Delta I^2 = \alpha^2 w^2\,\Delta f_0, \tag{20.16S}$$

linear

$$\Delta I^2 = \alpha^2 w\,\Delta f_0/8\pi. \tag{20.16L}$$

If the bandwidth Δf of the measuring instrument is very much less than Δf_0 a measurement of the d.c. current can be made to an accuracy $I_{dc} \pm (w\,\Delta F)^{\frac{1}{2}}$. For the square-law detector the fractional accuracy in I is therefore $(2\,\Delta F/\Delta f_0)^{\frac{1}{2}}$, while for the linear detector it is $(\Delta F/2\,\Delta f_0)^{\frac{1}{2}}$. In either case the accuracy with which w can be measured is $(2\,\Delta F/\Delta f_0)^{\frac{1}{2}}$.

20.5. Signal and noise in a square law detector

If the input to a square-law detector contains white noise w in Δf_0 and a unmodulated continuous wave signal $S \cos 2\pi f t_0$, the detector output may be calculated as in § 20.2. The calculation though elementary is tedious. The result for the d.c. term in the output is

$$I_{dc} = \frac{\alpha}{2}(S^2 + 2w\,\Delta f_0), \tag{20.17S}$$

so that signal and noise are detected independently.

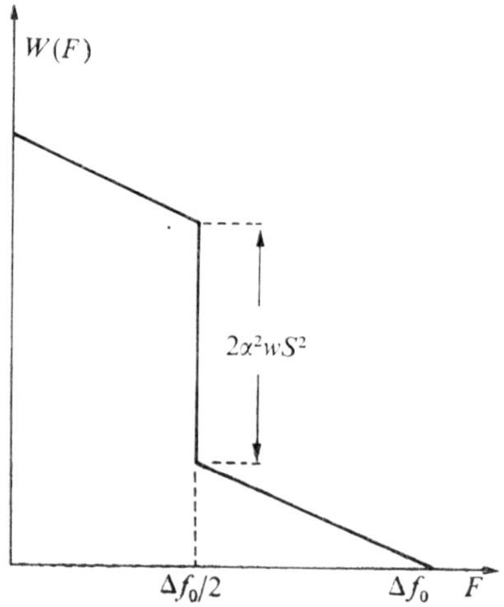

FIG. 20.4. Power spectrum $W(F)$ of the low-frequency output of a square-law detector with a white noise input in Δf_0 and a signal S at the centre of the pass-band.

The power spectrum is rather less simple since noise between d.c. and $F = \Delta f_0/2$ can arise either from beats between noise components or between noise and signal, whereas noise between $\Delta f_0/2$ and Δf_0 arises from noise beats alone. The results are

$$\left.\begin{aligned} W(F) &= 2\alpha^2 w\left\{S^2 + w\,\Delta f_0\left(1 - \frac{F}{\Delta f_0}\right)\right\}, \quad && 0 < F < \frac{\Delta f_0}{2}, \\ W(F) &= 2\alpha^2 w^2\,\Delta f_0\left(1 - \frac{F}{\Delta f_0}\right), && \frac{\Delta f_0}{2} < F < \Delta f_0. \end{aligned}\right\} \tag{20.18S}$$

This spectrum is illustrated in Fig. 20.4. The total low-frequency noise obtained by integrating eqn (20.18S) from d.c. to $F \geqslant \Delta f_0$ is

$$\Delta I^2 = \alpha^2 w\, \Delta f_0(S^2 + w\, \Delta f_0). \tag{20.19S}$$

We see that a strong signal increases the noise power output.

20.6. Signal and noise in a linear detector

Once again the results are more complicated. However, we can give simple asymptotic formulae valid (1) when the input signal to noise power ratio $x = S^2/2w\,\Delta f_0$ is small and (2) when it is large.

$x \leqslant 1$ (S/N small):

$$I_{\mathrm{dc}} = \alpha\left(\frac{w\,\Delta f_0}{2\pi}\right)^{\frac{1}{2}}\left(1+\frac{x}{2}\right), \tag{20.17La}$$

$$W(F) = \frac{\alpha^2 w}{4\pi}\left(1 - \frac{F}{\Delta f_0}\right), \tag{20.18La}$$

$$\Delta I^2 = \frac{\alpha^2}{8\pi} w\,\Delta f_0. \tag{20.19La}$$

$x \geqslant 1$ (S/N large):

$$I_{\mathrm{dc}} = \frac{\alpha S}{\pi}\left(1+\frac{1}{4x}\right), \tag{20.17Lb}$$

$$\left.\begin{aligned} W(F) &= \frac{2\alpha^2 w}{\pi^2}\left(1+\frac{1-F/\Delta f_0}{8x}\right), && 0 < F < \frac{\Delta f_0}{2}, \\ W(F) &= \frac{2\alpha^2 w}{\pi^2}\left(\frac{1-F/\Delta f_0}{8x}\right), && \frac{\Delta f_0}{2} < F < \Delta f_0, \end{aligned}\right\} \tag{20.18Lb}$$

$$\Delta I^2 = \frac{\alpha^2 w\,\Delta f_0}{\pi^2}\left(1+\frac{1}{8x}\right). \tag{20.19Lb}$$

We note from eqn (20.19Lb) that a large signal increases the noise by $8/\pi$ from the value for no signal. We also see from eqn (20.17La) that a small signal is reduced by the presence of noise. The part of the d.c. output attributable to the signal is

$$I_s = \tfrac{1}{2}\alpha\left(\frac{w\,\Delta f_0}{2\pi}\right)^{\frac{1}{2}} x = \frac{\alpha S^2}{4(2\pi w\,\Delta f_0)^{\frac{1}{2}}}.$$

We can understand this result qualitatively if we separate $R(t)$ into a noise part $n(t)$ and a signal part $S(t)$, for then

$$I_{\mathrm{dc}} \propto R(t) = +(n^2+S^2)^{\frac{1}{2}} \approx |n| + \frac{\frac{1}{2}S}{|n|},$$

and so

$$I_s \approx \frac{S^2}{|n|}.$$

20.7. Signal to noise ratio after square-law detection

In practice ΔF after detection is usually either very much less than Δf_0 or approximately equal to it. We consider first the case when $\Delta F \ll \Delta f_0$.

The signal to noise power ratio before detection is $x = S^2/2w\,\Delta f_0$. From eqn (20.17S) the output signal is

$$S_0 = I_{\text{dc}} - \alpha w \Delta f_0 = \alpha S^2/2,$$

while from eqn (10.18S) the noise output is

$$N_0^2 = 2\alpha^2 w\,\Delta F(S^2 + w\,\Delta f_0).$$

The output signal to noise power ratio is therefore

$$X = \frac{S_0^2}{N_0^2} = \frac{x^2\,\Delta f_0}{2(1+2x)\,\Delta F}. \tag{20.20S}$$

Thus for small signals the output ratio is proportional to the square of the input ratio and is unity when

$$x = \sqrt{(2\,\Delta F/\Delta f_0)}. \tag{20.21}$$

The input signal power is then

$$P_s = S^2/2 = xw\,\Delta f_0 = w\sqrt{(2\,\Delta F.\Delta f_0)}. \tag{20.22}$$

Sometimes this result is stated to imply that the effective bandwidth of an amplifier of bandwidth Δf_0 followed by a detector and a filter of bandwidth $\Delta F \ll \Delta f_0$ is

$$\Delta f_n = \sqrt{(2\,\Delta F.\Delta f_0)}. \tag{20.23}$$

for small signals. We should note. however, that neither is the output proportional to the input nor is the output signal to noise ratio proportional to that at the input.

For large signals $x \gg 1$, we have

$$X = \frac{x\,\Delta f_0}{4\,\Delta F}, \tag{20.24S}$$

and the two signal to noise ratios are proportional. It is legitimate therefore to talk about an effective noise bandwidth, which in this case is clearly

$$\Delta f_n = 4\,\Delta F. \tag{20.25S}$$

Thus, if a detector follows an amplifier of noise figure F_0, a signal power P_s yields an output signal to noise ratio

$$X = \left(\frac{S}{N}\right)_0^2 = \frac{P_s}{F_0 k T_0\,\Delta f_n}. \tag{20.26}$$

An increase δP_s in a large signal power P_s leads to an increase in signal output according to eqn (20.17S):

$$\delta S_0 = \alpha\,\delta P_s.$$

The noise output has an r.m.s. value from eqn (20.18S):

$$N_0 = \alpha\sqrt{(4P_s w\, \Delta F)} = \alpha\sqrt{(P_s w\, \Delta f_n)},$$

so that $\delta S_0 = N_0$ when

$$\delta P_s = \sqrt{(P_s w\, \Delta f_n)}. \tag{20.27S}$$

and not $2\sqrt{(P_s w\, \Delta f_n)}$ as for a linear system.

When $\Delta F \sim \Delta f_0$, eqns (10.17S) and (10.19S) yield

$$X = \frac{x^2}{1+2x}, \tag{20.28S}$$

so that unity output signal to noise ratio requires $x = 1+\sqrt{2}$ or

$$P_s = 2{\cdot}4w\, \Delta f_0. \tag{20.29S}$$

For large signals

$$X = x/2,$$

and so with the same reservations the effective bandwidth is

$$\Delta f_n = 2\, \Delta f_0. \tag{20.30S}$$

20.8. Signal to noise ratio after linear detection

We must now distinguish between cases when x is small and when x is large. For small values of x and $\Delta F \ll \Delta f_0$, eqns (20.17La) and (20.18La) yield

$x \ll 1, \quad \Delta F \ll \Delta f_0,$

$$X = \left(\frac{S}{N}\right)_0^2 = x^2 \frac{2\, \Delta f_0}{\Delta F}. \tag{20.20La}$$

Thus, for the linear detector too, the output ratio is proportional to the square of the input ratio, the signal power giving unity signal to noise ratio is again given by eqn (20.22), and the 'effective noise bandwidth' for small signals by eqn (20.23), i.e. $(2\, \Delta f_0\, \Delta F)^{\frac{1}{2}}$.

For large values of the input signal to noise ratio but with ΔF still small compared with Δf_0, we have from eqns (20.17Lb) and (20.18Lb)

$x \gg 1, \quad \Delta F \ll \Delta f_0,$

$$X = x\frac{\Delta f_0}{\Delta F}, \tag{20.24L}$$

and now, since X is proportional to x and the output signal is proportional to the input signal, we may define an effective noise bandwidth with no reservations. It is clearly

$$\Delta f_n = \Delta F. \tag{20.25L}$$

For small signals when $\Delta F \sim \Delta f_0$, we obtain from eqns (20.17La) and (20.19La),

$x \ll 1, \quad \Delta F \sim \Delta f_0,$

$$X = x^2, \tag{20.28La}$$

and for large signals from eqns (20.17Lb) and (20.19Lb)

$$x \gg 1, \qquad \Delta F \sim \Delta f_0, \qquad X = 2x. \tag{20.28Lb}$$

Neither of these formulae can be used to calculate the value of x that gives $X = 1$, but it is obviously near unity so that a signal

$$P_s \sim w\, \Delta f_0. \tag{20.29L}$$

will give a signal to noise ratio of unity.

From eqn (20.28Lb) the effective noise bandwidth for large signals is

$$\Delta f_n = \Delta f_0/2. \tag{20.30Lb}$$

The apparent paradox that the noise bandwidth is less than the signal bandwidth is explained when we realize that this result only applies to a strong signal at exactly f_0 which does not fully use the available bandwidth. Indeed the analysis in §§ 20.5–20.8 needs considerable modification if the signal, instead of being a continuous wave of a single frequency, contains several strong frequency components. This would, for example, be the case if, the signal were a modulated carrier wave.

20.9. Detector noise

When a detector forms the first stage of a system we are not only interested in how it modifies the effective bandwidth but also in noise generated in the detector and in its rectification efficiency, i.e. the ratio of the low-frequency output power to the radio-frequency input power. If this is low, noise generated in the low-frequency post-detection amplifier is enhanced in importance relative to the signal. Of course, we shall have also to distinguish between the threshold sensitivity, i.e. the weakest detectable signal, and the differential sensitivity to small changes in a strong signal. Detectors which have a poor threshold sensitivity may yet have a good differential sensitivity. There is no inherent reason why a detector should have a low threshold sensitivity; the photomultiplier which is, after all, a square-law detector, under ideal conditions, can detect single photons of energy $h\nu \sim 300\, kT$ and, were avalanche multiplication in a photo-diode more easily controlled, this limit might be pushed down by a further factor of 10 or so. The relative inefficiency of radio-frequency detectors is of practical rather than fundamental significance.

Detectors may be divided into two broad classes—those which rely on the non-linear transfer characteristic of an active device such as a transistor and those which use passive devices, notably diodes. Active devices are somewhat simpler to analyse, since usually the output circuit does not react back on the input, and the output noise level is such that noise generated in the output circuit and post-detection amplifier is insignificant.

If the current output I of an active device can be related to the input voltage v by

$$I = I_0 + g_m v + \gamma v^2, \tag{20.31}$$

and the device has an infinite input impedance, an input from a source of available signal power P_s and internal impedance R_s will result in a signal current

$$I_s = 4R_s\gamma P_s, \tag{20.32}$$

and, if the noise current in the output is expressed as

$$I_n^2 = 4nkTg_m\,\delta f, \tag{20.33}$$

the threshold signal is

$$P_{min} = \frac{(nkTg_m\,\delta f)^{\frac{1}{2}}}{2\gamma R_s}. \tag{20.34}$$

Consider, for example, an f.e.t. with a fixed (negative) gate bias $-V_0$, for which the drain–source saturated current is I_{dss}, so that

$$I = I_{dss}\left(1 - \frac{V_0}{V_p} + \frac{v}{V_p}\right)^2. \tag{20.35}$$

This gives

$$I_0 = I_{dss}\left(1 - \frac{V_0}{V_p}\right)^2,$$

$$g_m = \frac{2I_{dss}}{V_p}\left(1 - \frac{V_0}{V_p}\right),$$

and

$$\gamma = \frac{I_{dss}}{V_p^2}.$$

We then have

$$P_{min} = \frac{V_p}{R_s g_0}\left(\frac{g_m}{g_0}\right)^{\frac{1}{2}}(nkTg_0\,\delta f)^{\frac{1}{2}}, \tag{20.36}$$

where g_0 is the mutual conductance at zero bias. The noise ratio n is given by

$$n \sim 1 + f_0/f_2, \tag{20.37}$$

where f_0 is the flicker-noise corner-frequency and f_2 is the output frequency. If, for example, $g_0 = 5\text{ mA v}^{-1}$, $V_p = 2\text{V}$, $g_m/g_0 = \frac{1}{4}$, $T = 300\text{ K}$, and $R_s = 25\text{ k}\Omega$, corresponding to a reasonably high Q tuned input circuit, we obtain $P_{min} \sim 3{\cdot}6\times10^{-14}\ (n\,\delta f)^{\frac{1}{2}}$. If $f_0 = 5000$ Hz and the post-detection bandwidth extends from 10 Hz to 10^4 Hz, we have

$$n = \frac{1}{\delta f}\int_{10}^{10^4}\left(1 + \frac{5000}{f}\right)\mathrm{d}f \approx 6$$

and $P_{\min} \sim 10^{-11}$ W. This may be compared with thermal noise in the same bandwidth $\sim 4\times 10^{-17}$ W. This system is a factor $2{\cdot}5\times 10^{5}$ worse than ideal, and no practicable change in the parameters will make it much better.

The differential sensitivity is also given by (20.36), and this may be compared with differential sensitivity of a system consisting of a linear amplifier of noise figure F and the same bandwidth followed by a detector. This is

$$P_{\min} = 2(FkT\,\delta f P_{\mathrm{s}})^{\frac{1}{2}}, \tag{20.38}$$

and so the equivalent noise figure is

$$F = \frac{ng_{\mathrm{m}}}{4R_{\mathrm{s}}^{2}\gamma^{2}P_{\mathrm{s}}} = n\frac{g_{\mathrm{m}}}{g_{0}}\frac{1}{g_{0}R_{\mathrm{s}}}\frac{V_{\mathrm{p}}^{2}}{4P_{\mathrm{s}}R_{\mathrm{s}}} = n\frac{g_{\mathrm{m}}}{g_{0}}\frac{1}{g_{0}R_{\mathrm{s}}}\frac{V_{\mathrm{p}}^{2}}{v^{2}}. \tag{20.39}$$

With the numerical values given above, this gives $F = 1{\cdot}6\times 10^{-2}V_{\mathrm{p}}^{2}/v^{2}$; thus as the input voltage v approaches V_{p}, F becomes small, and eventually we should have to include, in the output noise, noise deriving from rectified input noise, which we have so far neglected, i.e. we replace (20.39) by

$$F = F_{0}+n\frac{g_{\mathrm{m}}}{g_{0}}\frac{1}{g_{0}R_{\mathrm{s}}}\frac{V_{\mathrm{p}}^{2}}{v^{2}}, \tag{20.40}$$

where F_0 is the noise figure of the transistor regarded as an amplifier. If the r.m.s. input signal voltage v exceeds about $V_{\mathrm{p}}/5$ the effective noise figure is not far from F_0. We see that this system, which is a very bad threshold detector for weak signals, is a good detector for small changes in a strong signal.

A somewhat more complicated active detector is shown in Fig. 20.5. This is an exceedingly poor threshold detector but, as we shall see, a good linear differential detector. If $g_0 R_1$ is large a relatively small input voltage v switches the current from T_1 to T_2, and the current through T_1 as a function of v

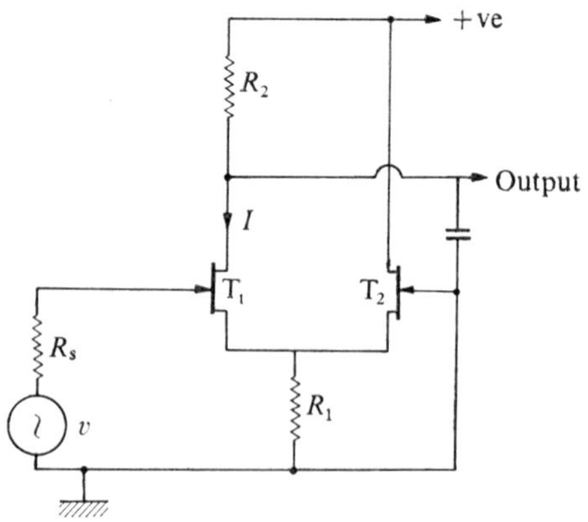

FIG. 20.5. Envelope detector.

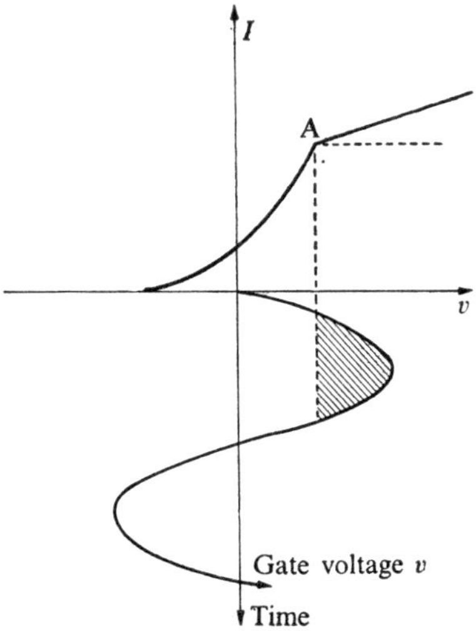

FIG. 20.6 Envelope detection.

behaves as shown in Fig. 20.6. The transistor T_2 is turned off at the point A and, for more positive voltages, I increases linearly with v as v/R_1. For large enough amplitudes the rectified current is

$$I_s = v/\pi R_1. \tag{20.41}$$

Provided that $g_m R_1$ is greater than n (the noise ratio in eqn (20.33)) the low-frequency noise in T_1, which is on for half the time, is

$$I_n^2 = \frac{2kT\,\delta f}{R_1}. \tag{20.42}$$

The change δv in v that results in unity signal to noise ratio is therefore

$$\delta v = \pi(2R_1 kT\,\delta f)^{\frac{1}{2}}. \tag{20.43}$$

This may be compared with the r.m.s. noise voltage due to the source

$$\delta v_s = (4F_0 R_s kT\,\delta f)^{\frac{1}{2}}, \tag{20.44}$$

and so, if $F_0 R_s \gg (\pi^2/2)R_1$, this will be the dominant term and the detector will be almost ideal as a differential system, as well as being a linear detector.

The value of R_1 required can be determined from the relations

$$g_0 R_1 = \left(\frac{g_0}{g_m}\right)^2 - \frac{g_0}{g_m},$$

$$g_m R_1 = \frac{g_0}{g_m} - 1,$$

where g_m is the mutual conductance in the quiescent condition and g_0 the mutual conductance at zero bias. If, for example, we make $g_m/g_0 = \frac{1}{10}$, then $g_m R_1 = 9$ and $g_0 R_1 = 99$. The input voltage required to switch the transistors is $v = V_p(g_m/g_0)$ and so in this case is $V_p/10$. The required input drive is therefore somewhat larger than this, say $V_p/2$. Finally, the source resistance must satisfy $R_s > \frac{1}{2}\pi^2 R_1 \approx 500/g_0$, so that, if $g_0 = 5$ mA v^{-1}, we require $R_s > 100$ kΩ. These values are somewhat difficult to realize in practice, nevertheless practical circuits based on Fig. 20.5 can be constructed which, in the frequency range up to about 30 MHz, have effective noise figures less than 10 for inputs of around 1 V amplitude.

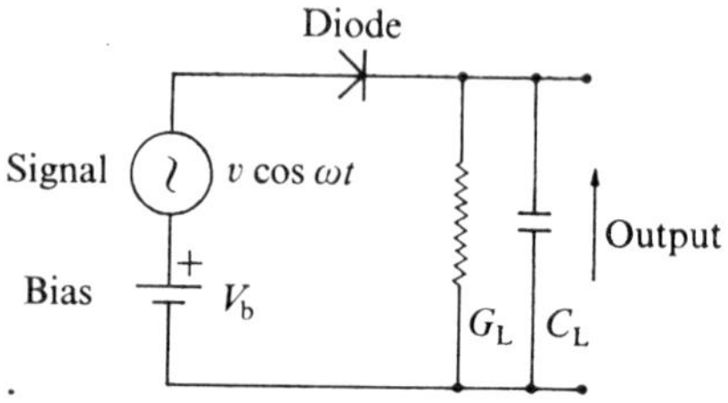

FIG. 20.7. Diode detector.

The basic diode detector circuit is shown in Fig. 20.7, and in this case we have to consider both noise generated in the diode and Johnson noise associated with the load conductance G_L. If the diode capacitance is neglected, ωC_L is assumed to be large, and the diode characteristic is represented as

$$I(V) = I_0\{\exp(\beta V) - 1\}, \tag{20.45}$$

then, for a bias current I, we find that an input power P results in a rectified output voltage

$$v_s = \frac{\frac{1}{2}\beta P}{G_L + \beta(I + I_0)}. \tag{20.46}$$

The mean square output noise voltage is

$$v_n^2 = \frac{2e(I+2I_0)\,\delta f + 4kTG_L\,\delta f}{\{G_L + \beta(I+I_0)\}^2}, \tag{20.47}$$

and the threshold signal power is

$$P_{min} = \frac{\{8e(I+2I_0)\,\delta f + 16kTG_L\,\delta f\}^{\frac{1}{2}}}{\beta}. \tag{20.48}$$

For an ideal diode we can ignore I_0 and replace β by e/kT and if, at the same time, we choose I so that the required source conductance G_s is matched by the diode conductance βI, we have

$$P_{\min} = \frac{kT}{e}\{(8G_s+16G_L)kT\,\delta f\}^{\frac{1}{2}}, \tag{20.49}$$

so that

$$\frac{P_{\min}}{kT\,\delta f} = \left(\frac{24kT}{e^2R\,\delta f}\right)^{\frac{1}{2}}, \tag{20.50}$$

where we have set $R = G_s^{-1} = G_2^{-1}$. If, for example, $T = 300$ K, $R = 1000\ \Omega$, and $\delta f = 10^7$ Hz this ratio is about 10^4 and clearly, with reasonable component values, a diode makes a poor threshold detector.

The effective noise figure as a differential detector is obtained as

$$F = \frac{P_{\min}^2}{4kTP_s\,\delta f} = \frac{24}{4R_sP_s}\left(\frac{kT}{e}\right)^2 = 6\left(\frac{kT}{ev}\right)^2, \tag{20.51}$$

and becomes tolerable as the input voltage approaches $kT/e = 25$ mV. At this level, however, the power series expansion of $\exp \beta V$ becomes insufficient, and the detector enters the linear regime.

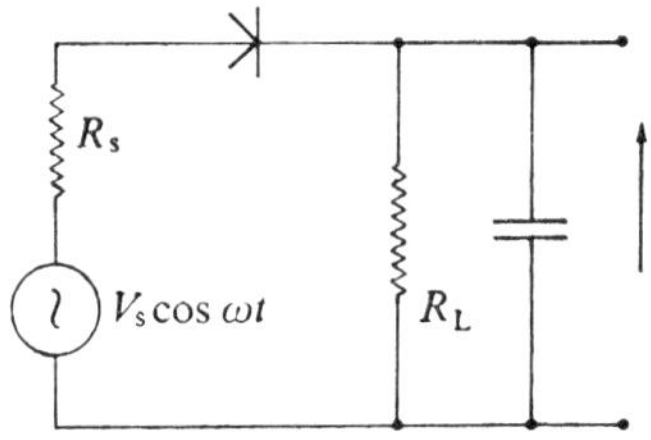

FIG. 20.8. Linear detection with a diode.

In Fig. 20.8 the diode represents a perfect switch which closes only when $V_s \cos \omega t > V_0$. If we let $\theta = \cos^{-1} V_0/V$, the power delivered by the source averaged over a cycle is

$$\frac{V_0^2}{R_L} = P = \frac{1}{2\pi R_s}\int_{-\theta}^{\theta} V_0(V_s \cos \omega t - V_0)\, \mathrm{d}(\omega t),$$

which gives

$$\frac{V_0^2}{R_L} = \frac{V_0}{\pi R_s}(V_s \sin \theta - V_0\theta),$$

so that

$$\frac{\pi R_s}{R_L} = \tan \theta - \theta$$

and the detection efficiency is

$$\alpha = \frac{4PR_s}{V_s^2} = \frac{4}{\pi}\cos \theta(\sin \theta - \theta \cos \theta).$$

This gives a maximum efficiency of about $\frac{1}{4}$ when θ is slightly greater than 1 radian and $R_s \sim 0{\cdot}4R_L$. For a full wave rectifier the efficiency is approximately $\frac{1}{2}$. Thus with a full wave rectifier the effective noise figure as a differential detector approaches $2F_0$, where F_0 is the noise figure of the post-detection amplifier.

The above considerations apply qualitatively to microwave crystal detectors, whose rectification efficiency is often expressed empirically in terms of the input power P as

$$\alpha = \frac{1}{2}\frac{\eta P}{1+\mu P}. \tag{20.52}$$

Typical values of η and μ are $10^3\ \text{W}^{-1}$ and $2\times10^3\ \text{W}^{-1}$. In this case we cannot neglect flicker noise generated by the diode, and this is usually expressed in terms of a noise ratio n, so that the available noise output power is $nkT\,\delta f$. The ratio n increases with output power and can be expressed as

$$n = 1+\gamma P_0/f, \tag{20.53}$$

where P_0 is the output power αP. A typical value for γ is $10^{12}\ \text{J}^{-1}$. We then find for the effective noise figure as a differential detector

$$F = \left(\frac{1+\mu P_i}{2+\mu P_i}\right)^2\left\{\frac{2(1+\mu P_i)}{\eta P_i}+\frac{\gamma P_i}{f}\right\}. \tag{20.54}$$

This has a minimum value when

$$P_i \sim (2f/\eta\gamma)^{\frac{1}{2}}. \tag{20.55}$$

If $f = 1$ kHz, $\eta = 10^3\ \text{W}^{-1}$, and $\gamma = 10^{12}\ \text{J}^{-1}$ this corresponds to $1{\cdot}4\ \mu$W, well into the square-law regime. The corresponding minimum value of F is

$$F_{\min} \sim (\gamma F_0/2\eta f)^{\frac{1}{2}}, \tag{20.56}$$

which for these numerical values, is 700.

As a threshold detector we have

$$P_{\min} \sim \left(\frac{2kT\,\delta f}{\eta}\right)^{\frac{1}{2}}. \tag{20.57}$$

Thus with $\eta = 10^3\ \text{W}^{-1}$ and $\delta f = 1$ MHz we obtain $P_{\min} \sim 3\times10^{-9}$ W.

It should be clear from this discussion that practical detectors are always very insensitive as threshold detectors and that, in every case, low noise detection approaching the theoretical limits will only be possible with substantial radio-frequency amplification before the detector. On the other hand, the loss in sensitivity in the differential detection of a small change in a strong carrier is much less marked. This is particularly important in physical instruments such as radio-frequency spectrometers, where the signal is present as a small modulation of a carrier.

21 Photomultipliers and photo-diodes

21.1. Introduction

ALTHOUGH both photomultipliers and photo-diodes are square-law devices, giving an output current proportional to input power, they differ from radio-frequency square-law detectors in that the energy quantum $h\nu$ of the incoming radiation is much greater than the thermal energy kT associated with the detector and, in some circumstances, the fluctuations in the output current are determined almost entirely by the fluctuations in the incident radiant energy flux. Their performance is intermediate between radio-frequency square-law detectors and the particle counters of nuclear physics which, because the incoming quanta are so large, not only count these quanta but also can furnish information about their energy.

In both photo-detectors the primary process is the generation of a mobile electron by the absorption of a photon. In some cases, e.g. a silicon photo-diode near 9000 Å, the quantum efficiency of this process is near unity, and on average each incident photon releases one electron; in other cases. e.g. photo-emissive detectors near the red end of the visible spectrum, the quantum efficiency is much less, and on average only one photon in a hundred or so releases an electron.

In an ideal photo-detector not only is the quantum efficiency of the primary process unity but also it is possible to amplify the resulting electron current without adding further noise. Real photo-detectors fall short in one or the other of these respects and may also exhibit a fluctuating output current even in the absence of any light input. The relative importance of these factors varies from device to device. In photo-diodes the primary quantum efficiency may approach unity and the background current be negligible, but, even if avalanche multiplication is used, the resulting gain is too low to make the noise of the succeeding amplifier negligible, except at high photon fluxes. By contrast, the photomultiplier, through secondary-emission multiplication, has enough gain but usually suffers from a very low primary quantum efficiency and, unless the photo-cathode is cooled, from an appreciable background or dark current.

The fluctuations in the incident radiation depend on the nature of the source and, if the detector is sufficiently sensitive, a record of these fluctuations, or the correlations between the fluctuations in two different detectors responding to the same source, can be used to deduce some of the properties of the source. For our purposes, however, it will be sufficient to regard the incident radiation as a stream of photons displaying shot noise. Thus if, in some period τ, the average number of incident photons is N, individual observations will fluctuate about N with a root mean square deviation $N^{\frac{1}{2}}$. If the incident power is P, photons arrive at a mean rate $P/h\nu$ and so, with unity quantum efficiency, the primary current is $I = eP/h\nu$, displaying shot noise fluctuations of power spectrum $2eI = 2e^2P/h\nu$. If this current could be amplified noiselessly by a system, with a response time less than $h\nu/P$, the resulting device would detect single photons. Now suppose that each primary electron is fed to an amplifier whose input admittance consists of a capacitance C in parallel with a resistance R. If R is at a temperature T the charge fluctuations associated with C have a mean square value kTC and, unless this is less than e^2, we have no hope of detecting single electrons or photons. Since $e^2 \sim 2{\cdot}5\times10^{-38}$ C^2 while, even if C has the very low value of 1 pF and T is 1 K, $kTC = 1{\cdot}4\times10^{-35}$ C^2, we see that we have no hope of detecting single photons unless some form of amplification is inherent in the detector structure itself. On the other hand, suppose that we wish to observe small changes in the intensity of a strong beam of radiation, then we have to compare the power spectrum $2eI$ of the current fluctuations in the primary current with the power spectrum $4kT/R$ associated with R. Provided that $eIR > 2kT$, which requires $20IR > 1$ (if $T = 300$ K), the fluctuations in the amplifier input will be mainly due to fluctuations in the radiation. If, for example, $R = 10^5\ \Omega$ we require f greater than about 2μA which, with photons of 2eV energy and unit quantum efficiency, implies an incident photon flux of about 10^{-6} W, or some 10^{13} photons per second. The input capacitance of the amplifier does not enter directly into this calculation. but it does limit the value of R (for a given bandwidth δf) to $1/(2\pi C\,\delta f)$.

21.2. Quantum efficiency

We now calculate the fluctuations in the primary photo-electron current when, on the average, one incident photon results in $\eta < 1$ electrons. Consider first the case of an interval τ in which exactly k photons arrive, then, if α is the probability that a photon results in an electron, the probability that k photons result in m electrons is ${}^kC_m\alpha^m(1-\alpha)^{k-m}$ and the average number of electrons emitted in τ is

$$\langle m\rangle_k = \sum_{m=0}^{k} m\,{}^kC_m\alpha^m(1-\alpha)^{k-m} = \alpha k. \tag{21.1}$$

We have also

$$\langle m(m-1)\rangle_k = k(k-1)\alpha^2 = \langle m\rangle^2 - k\alpha^2,$$

so that

$$\langle\Delta m^2\rangle = \langle m^2\rangle - \langle m\rangle^2 = \langle m\rangle - k\alpha^2 = k\alpha(1-\alpha) = (1-\alpha)\langle m\rangle. \tag{21.2}$$

We see from (21.1) that α is equal to η the quantum efficiency and from (21.2) that if $\alpha = \eta = 1$ there are no additional fluctuations due to the photo-emission process, whereas if $\alpha = \eta \ll 1$ the electron current will display full shot noise, since then $\langle \Delta m^2 \rangle = \langle m \rangle$. Now we consider the case where the photons arrive at random times; we divide τ into N infinitesimal intervals t, so that the probability p that a single photon arrives in an interval t is small and the probability that k photons arrive in τ is $P(k, \tau) = {}^N C_k p^k (1-p)^{N-k}$. We have for the probability that k photons arrive and result in m electrons,

$$P(m, k, \tau) = {}^N C_k p^k (1-p)^{N-k}\, {}^k C_m \alpha^m (1-\alpha)^{k-m}.$$

Thus

$$\langle m \rangle = \sum_{k=0}^{N} {}^N C_k p^k (1-p)^{N-k} \sum_{m=0}^{k} m\, {}^k C_m \alpha^m (1-\alpha)^{k-m} =$$

$$= \sum_{k=0}^{N} \alpha k\, {}^N C_k p^k (1-p)^{N-k} = \alpha N p.$$

Similarly, $$\langle m(m-1) \rangle = \alpha^2 p^2 N(N-1) = \langle m \rangle^2 (1-1/N) \approx \langle m \rangle^2.$$

and

$$\langle \Delta m^2 \rangle = \langle m \rangle. \tag{21.3}$$

In this case the emitted electron current displays full shot noise whatever the value of the quantum efficiency $\alpha = \eta$. If $\eta < 1$ the r.m.s. fluctuations in the electron current are less than the fluctuations in the radiation by $\eta^{\frac{1}{2}}$ but the signal current is less by η, and so the signal to noise ratio is worse by a factor $\eta^{-\frac{1}{2}}$. Whatever the nature of the amplification process to which the electron current is subsequently subjected, this initial loss in signal to noise ratio cannot be recovered.

21.3. Secondary-emission multiplication

We have seen in the introduction to this chapter that we have no chance of achieving single-photon detection, or photon counting, if the primary photo-electron current is fed directly to a conventional amplifier. In a photo-multiplier tube the emitted photo-electrons are first accelerated to about 100 eV energy and then impinge on a secondary emission surface. Secondary electrons from this surface are then further multiplied in a sequence of upwards of 5, but usually 10 or more, stages. We must discuss now the additional fluctuations introduced in this process.

If the primary current displays full shot noise, so that the number of primary electrons arriving in an interval τ has a Poisson distribution about its mean number N, and this current strikes a secondary emission surface, releasing on average $\langle \sigma \rangle$ electrons for every incident electron, then the mean number of secondary electrons emitted is $\langle n \rangle = N\langle \sigma \rangle$ and the mean square deviation in this number is $\langle \Delta n^2 \rangle = N\langle \sigma \rangle^2$, which we can also write as $\langle \Delta n^2 \rangle = N\langle \sigma \rangle^2 + N\langle \Delta \sigma^2 \rangle$. (See e.g. Feller 1957, Chap. 12.) The emission of secondary electrons is the result of a large number of random processes within the surface of the electrode, and we may assume that σ has a Poisson distribution about $\langle\sigma\rangle$. Thus

$$\langle\Delta n^2\rangle = N\langle\sigma\rangle^2+N\langle\sigma\rangle. \tag{21.4}$$

If the mean input current is I_0, the mean output current after one stage of multiplication is $I_1 = \sigma I_0$, where now σ denotes the average multiplication factor, i.e. $\langle\sigma\rangle$, and the power spectrum of the output current is

$$w_1 = 2eI_0\sigma^2+2eI_1 = \sigma(\sigma+1)2eI_0 = (\sigma+1)2eI_1. \tag{21.5}$$

After k stages the square of the output current is

$$I_k^2 = \sigma^{2k}I_0^2, \tag{21.6}$$

and the power spectrum of the output fluctuations is

$$w_k = 2eI_0(\sigma^{2k}+\sigma^{2k-1}+\ldots+\sigma^k) = 2eI_0\sigma^{2k}\frac{1-\sigma^{-(k+1)}}{1-\sigma^{-1}}. \tag{21.7}$$

If σ is appreciably greater than unity and k is large enough, this is approximately

$$w_k = 2eI_0\sigma^{2k}\frac{\sigma}{\sigma-1} \tag{21.8}$$

and we see, by comparing w_k/I_k^2 with the initial ratio $w_0/I_0 = 2e$, that the mean square signal to noise ratio has deteriorated by a factor F^{-1}, where

$$F = \frac{\sigma}{\sigma-1}, \tag{21.9}$$

a result first derived by Shockley and Pierce (1938). If, for example, $\sigma = 3$ we have $F = 1{\cdot}5$. The over-all current gain of the multiplier is $\Gamma = \sigma^k$, and the average size of the individual output pulses is Γe. If these pulses drive an amplifier of input capacitance C with an input conductance at T, single pulses will be detectable if

$$\Gamma^2e^2 > kTC. \tag{21.10}$$

Thus if $C = 10$ pF and $T = 300$ K we require $\Gamma > 10^3$. If $\sigma = 3$ this requires only 7 or 8 stages of secondary-emission multiplication. Alternatively, if the multiplier output current develops a voltage across a load resistance R we require

$$\Gamma^2eI_0 > 2kT/R, \tag{21.11}$$

and if with 12 stages we have $\Gamma = 10^5$ and $T = 300$ K, we only require $I_0R > 5\times10^{-12}$ V. Thus, if $R = 1000\ \Omega$, $I_0 > 5\times10^{-15}$ A or approximately 3×10^4 electrons per second. With $C = 10$ pF the bandwidth of the input stage would be 16×10^6 Hz, and clearly this system would produce distinct pulses as each primary electron entered the system.

Although secondary-emission multipliers are most usually employed in conjunction with a photo-emissive surface in a photo-multiplier tube they may also be used to amplify the current due to any incident stream of charged particles, e.g. the beam current of ions in a mass spectrometer. In this case the average secondary-emission multiplication factor σ_1 of the first surface may be different from that of succeeding surfaces, since this surface converts ions rather than electrons to secondary electrons. In this case it is easy to see that with a large number of stages

$$F = 1+\frac{\sigma}{\sigma_1(\sigma-1)}. \qquad (21.12)$$

Thus if σ_1 is large the value of σ in the succeeding stages is not critical, for example, if $\sigma_1 = 5$ and $\sigma = 2{\cdot}5$ we have $F = 1{\cdot}33$.

We end this section by reiterating that the main reason for using secondary-emission multiplication is to bring the signal and noise associated with the primary photo-electrons up to a level exceeding the input noise of the succeeding amplifier. It will be obvious from the numerical examples quoted in connection with eqns (21.10) and (21.11) that this is not difficult.

21.4. Photomultipliers

In a photomultiplier the incident radiation falls on a photo-emissive surface and, if the photon quantum $h\nu$ exceeds the work function W of the surface, electrons will be emitted which may be accelerated to impinge on the first dynode of a secondary-emission multiplier. The electron yield, or quantum efficiency η, depends rather critically on the difference $h\nu - W$ and is small if this difference is small. Since reproducible photo-emissive surfaces with W below about 1·5 eV do not exist, the quantum efficiency is very low, (of the order of 10^{-2} or less) for red light, whereas it may be greater than 10^{-1} for blue light. With a quantum efficiency η, the mean square signal to noise ratio in the primary photo-electron current is reduced by η, and so the overall noise figure of a photomultiplier is

$$F = \frac{1}{\eta}\frac{\sigma}{\sigma-1}, \qquad (21.13)$$

where σ is the dynode secondary-emission multiplication factor.

Because it is necessary to use a photo-emissive surface of low work function, thermionic emission from the surface cannot be neglected even at room temperature. We can make a rough estimate of its magnitude from the Dushman formula: $I = 10\,AT^2 \exp(-W/kT)$, where A is the cathode area in square centimetres. With $W = 1{\cdot}5$ eV and $A = 10$ cm^2 this gives 10^{-18} A or about 6 electrons per second at 300 K. With a quantum efficiency of 10^{-2} this would be equivalent to an input flux of 600 photons per second. Since the surfaces used to obtain an appreciable quantum efficiency are rather complex, the actual values of the dark current depart quite considerably from the predictions of the Dushman formula, which only applies to clean metal surfaces, but, in general, the dark current is the limiting factor in the sensitivity of photomultipliers to small light fluxes at room temperature. It can be reduced to negligible proportions by cooling the device.

21.5. Photo-diodes

If radiation, with $h\nu$ greater than the bandgap, is incident on a semiconductor such as silicon, the absorption of a photon may promote an electron from the valence band to the conduction band. If the semiconductor is initially of very low conductivity and the electron–hole recombination rate is low, the concentration of mobile carriers, electrons and holes, may make an appreciable difference to the conductivity, and the system will operate as a photo-conductive detector. The photo-diode, however, operates on a different principle and consists of a p–n junction diode to which reverse bias is applied. This bias completely inhibits the passage of electrons (the majority carriers in the n material) from the n side to the p side, and similarly it inhibits the passage of holes from p to n. On the other hand, any minority carriers (holes in the n material or electrons in the p material) are swept across the depletion layer at the junction and contribute to the current in the external leads. The current due to thermal generation of minority carriers constitutes the normal reverse current, or dark current, of the diode but, if photon absorption occurs anywhere near the junction, either the hole generated in the n material or the electron in the p material will contribute an additional component to the current. In suitable circumstances photo-diodes can be constructed with almost unity quantum efficiency. All the photons are absorbed near the junction and each photon yields a hole–electron pair. This is especially the case if $h\nu$ is only just greater than E_g the energy gap and, by contrast with photo-emissive surfaces, the quantum efficiency tends to fall with further increase in $h\nu$.

If the quantum efficiency is unity each incident photon yields one electron in the external circuit, and the resulting current must be amplified by conventional means. Since both the amplifier and the diode have an appreciable capacitance this means that we cannot use a photo-diode to detect single photons. On the other hand, if the incident photon flux is high and yields a current whose shot noise swamps the noise at the amplifier input, we can observe fluctuations which mainly originate in the incoming radiation. For example, if the incoming power is 15×10^{-6} W of 1·5 eV photons, the resulting current is 10^{-5} A and, with a load resistance of 10^4 Ω at 300 K, the shot-noise fluctuations exceed the current noise associated with the resistance by a factor of 2. In this sense a photo-diode operating at high input levels approaches the performance of an ideal detector, although it can never be used to count individual photons.

In the avalanche photo-diode the reverse bias across the junction is increased to a point where each primary hole–electron pair produces an avalanche of further pairs by impact. It is possible to obtain stable current gains of the order of 100 or more by this process, which, if it were noiseless, would improve the over-all sensitivity of the diode–amplifier combination by the same factor. The process is, however, in some ways analogous to secondary-emission multiplication with a low average multiplication factor per impact, and so leads to an increase in noise and a degradation of the signal to noise ratio. Nevertheless, the over-all signal to noise ratio of the diode–amplifier combination may still be very appreciably improved at low levels by using an avalanche diode. At levels where the simple photo-diode already gives near-ideal response no advantage, and indeed some deterioration, in the signal to noise ratio would result from the use of an avalanche diode.

The main advantages of a photo-diode over a photomultiplier are the higher quantum efficiency of the primary photo-electric process and its capacity to handle higher input powers without overloading. At high input levels, generally above 1 μ W, it can approach ideal sensitivity but, compared with a photomultiplier, it is relatively insensitive at low levels.

21.6. The photo-diode as a mixer

At high input levels a photo-diode can be regarded as an ideal square-law device and, if it is simultaneously irradiated with two incident beams of light of frequencies ν and $\nu+\delta\nu$ which are spatially coherent over the active region of the diode, it will generate a current at the beat frequency $\delta\nu$. If an incident power P yields a current $I=\alpha P$ and P_1 is the power incident at ν and P_2 at $\nu+\delta\nu$, the mean current will be $\alpha(P_1+P_2)$ and the r.m.s. amplitude of the difference frequency current will be $\alpha(2P_1P_2)^{\frac{1}{2}}$. Thus the output signal to noise ratio for an output bandwidth δf will be

$$\left(\frac{S}{N}\right)^2=\frac{2\alpha^2P_1P_2}{2e\alpha(P_1+P_2)\,\delta f}=\frac{\eta P_1P_2}{h\nu(P_1+P_2)\,\delta f}, \tag{21.14}$$

where we have expressed α as $\eta e/h\nu$ in terms of the quantum efficiency η.

In a mixer one of the powers, say P_2, will represent the local oscillator input and be much larger than the signal input P_1. If the diode load resistor is R and $(e\eta/h\nu)P_2 \gg 2kT/R$, noise from R can be disregarded, and thus (21.14) gives the actual signal to noise ratio of the whole system. With $P_2 \gg P_1$ we obtain

$$\left(\frac{S}{N}\right)^2=\frac{\eta P_1}{h\nu\,\delta f}, \tag{21.15}$$

and, when $\eta = 1$, this indicates that a mixer with a strong local-oscillator input gives unity signal to noise ratio with an input of 1 photon per time interval $1/\delta f$, i.e. it approximates the ideal sensitivity.

This property is not of much use if the sources P_1 and P_2 are incoherent in time, since the bandwidth δf will have to accommodate the very large spread which may occur in the values of $\delta\nu$ but, if both P_1 and P_2 are derived from stable single-frequency laser sources, a photo-diode mixer can be used to construct an extremely sensitive optical superheterodyne receiver.

22 Applications

22.1. Noise-figure measurements

THE usual way of measuring the noise figure or noise temperature of an amplifier is by observing the increase in the output noise when additional calibrated white noise is introduced at the input. This yields the average noise figure without necessitating a measurement of the amplifier gain and frequency response or needing a calibrated low-level signal source.

The experimental technique depends on the frequency and the expected value of the noise temperature T_n. If T_n is less than about 500 K it is usually enough to connect a dummy source of the correct impedance to the amplifier input and to plot the mean square output noise, or the output of a square-law detector, against the source temperature. If the detector is accurately square-law the plot will be a straight line with an intercept on the T axis at $-T_n$. Deviations from a straight line indicate that the detector is not square-law. If a square-law detector is inconvenient a linear detector may be used and its output plotted against $T^{\frac{1}{2}}$.

For higher values of T_n and at frequencies up to about 200 MHz, a temperature-limited thermionic diode provides a convenient variable and calculable source of noise. A typical circuit is shown in Fig. 22.1. The filter chokes L, the bypass capacitors C', and the coupling capacitor C'' depend on the frequency range, and it may also be necessary to add further circuit elements to tune out the diode capacitance, so that R_s in Fig. 22.1 corresponds to the actual source resistance. The diode current I is varied by varying the heater voltage, and the equivalent noise input power spectrum is

$$W = 2eIR_s^2+4k(T+T_n)R_s = 2eIR_s^2+4FkTR_s. \tag{22.1}$$

A plot of the mean square noise output against I gives an intercept at

$$I = -\frac{2FkT}{eR_s}. \tag{22.2}$$

If, for example, $F = 10$ and $R_s = 100\ \Omega$ the currents required are in the range 0–20 mA. Noise diodes for this purpose are made by many manufacturers. A typical example is the G.E.C. type CV 2398, which has a maximum anode current of 70 mA and delivers a white noise output between 1 kHz and 200 MHz. With a correction for transit-time effects it can be used up to 500 MHz. Because few detectors are either accurately square-law or accurately linear, it is safer to use the calibrated attenuator between the amplifier and the detector to keep the detector reading constant as I is varied.

At microwave frequencies coaxial line noise diodes are occasionally used, but more often the source is a gas discharge tube (Mumford, 1949). The electrons in the positive column of a gas discharge are at a high temperature $T_e \sim 15\,000$ K $\sim 50\ T_0$, which, in commercial tubes, is relatively insensitive to the discharge current. If a discharge tube is coupled to a transmission line

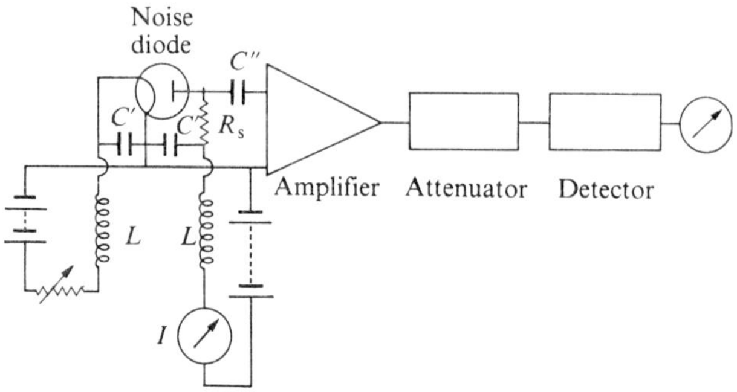

FIG. 22.1. Use of a temperature-limited noise diode to measure the noise figure of an amplifier.

or wave guide so that it presents a matched load to the system, the power spectrum of the noise output from the line is kT_e. This is at least enough to double the noise output of most amplifiers. If the source is coupled to the amplifier via an attenuation α, at a temperature T, the equivalent input noise power spectrum is

$$W = \alpha kT_e + (1-\alpha)kT + kT_n = \alpha k(T_e - T) + FkT, \qquad (22.3)$$

and a plot of the mean square noise output against α gives an intercept at

$$\alpha = -\frac{FT}{T_e - T} \sim -\frac{F}{50}.$$

Absolute noise measurements yielding T_n to an accuracy of better than ± 25 per cent are exceedingly difficult. When the noise performance of equipment is quoted to higher accuracy this is at least as likely to indicate an uncritical attitude to measurements as superior performance.

22.2. Aerial noise

When a transmitting aerial is excited by a signal source, a fraction η of the delivered power is radiated and a fraction $1-\eta$ is dissipated as heat in ohmic loss in the aerial and the matching network. If the aerial impedance is

$Z = jX + R$ and R is expressed as the sum of an ohmic term R_a and a radiation term R_r, the efficiency is

$$\eta = R_r/(R_r + R_a). \tag{22.4}$$

Unless the aerial dimensions bear some special relation to the wavelength λ the reactance X will be much larger than R_r and an appreciable part, if not all, of the ohmic resistance R_a will be associated with the network (of reactance $-X$) needed to match the aerial. This is especially the case with a short dipole of length $l \ll \lambda$. If the r.m.s. current is I the power radiated is

$$P = I^2 R_r = \frac{1}{6\pi}\left(\frac{\mu_0}{\epsilon_0}\right)^{\frac{1}{2}}\left(\frac{2\pi l}{\lambda}\right)^2 I^2, \tag{22.5}$$

but the current returns as displacement current though the capacitance C between the two limbs of the dipole. Thus in the equivalent circuit R_r appears in series with C. The major part of the power supplied to drive the aerial will be dissipated in the inductance used to cancel C, unless its quality factor exceeds

$$Q = 1/\omega C R_r. \tag{22.6}$$

This is proportional to $(\lambda/l)^3$. Only if the dipole is an integral number of half-wavelengths long is X zero and the over-all efficiency near unity.

The polar diagram of a transmitting aerial is described by its polar gain $g(\theta, \phi)$ defined so that the fraction dP of the total radiated power P going into a solid angle $d\Omega$ at (θ, ϕ) is

$$dP = P\, g(\theta, \phi)\, d\Omega/4\pi. \tag{22.7}$$

If the aerial is used for reception, its cross-section or receiving area $A(\theta, \phi)$ is defined so that an incident flux S Wm^{-2} from the direction (θ, ϕ) leads to a power

$$P = SA(\theta, \phi). \tag{22.8}$$

at its terminals; however, only a fraction η of this power can be delivered to an external load.

We now show that

$$A(\theta, \phi) = \lambda^2 g(\theta, \phi)/4\pi. \tag{22.9}$$

Consider an aerial matched to a resistive load with all its ohmic loss at T and which therefore radiates a total power kT per unit frequency interval. If a black body which presents an area a to the aerial is placed at a distance r, the power it receives from the aerial is $kTg(\theta, \phi)a/4\pi r^2$. If it is also at a temperature T the power it radiates per unit frequency interval in one polarization mode into a solid angle $d\Omega$ is, according to the Rayleigh–Jeans law, $akT\, d\Omega/\lambda^2$, and so the power received by the aerial is $akTA(\theta, \phi)/\lambda^2 r^2$. These two powers must be equal, and this yields (22.9). Clearly, for an isotropic aerial the effective area is $\lambda^2/4\pi$ in all directions. For a short dipole the gain at an angle θ to the axis is

$$g = \tfrac{3}{2}\sin^2\theta \tag{22.10}$$

and its effective area in this direction is

$$A = \frac{3\lambda^2}{8\pi} \sin^2\theta. \tag{22.11}$$

At right-angles to the axis A has its maximum value

$$A = 3\lambda^2/8\pi. \tag{22.12}$$

Note that A is independent of the length of the dipole as long as it is much less than a wavelength. The length only alters R_r and the efficiency η.

If an aerial of efficiency η, all of whose parts are at a temperature T_0, has a gain $g(\theta, \phi)$ and receives noise power from a region where the absorbing medium in a direction (θ, ϕ) is at temperature $T(\theta, \phi)$, the available noise power at the teminals is kT_a where

$$T_a = (1-\eta)T_0 + \eta \int g(\theta, \phi) T(\theta, \phi)\, d\Omega. \tag{22.13}$$

It is convenient to define an effective sky temperature T_s and to write this as

$$T_a = (1-\eta)T_0 + \eta T_s. \tag{22.14}$$

The signal energy flux S from a direction (θ, ϕ) required to give unity signal to noise ratio at the terminals is

$$S_{\min} = \frac{4\pi}{\lambda^2 g(\theta, \phi)} \left\{ kT_s + kT_0\left(\frac{1}{\eta} - 1\right) + \frac{kT_n}{\eta} \right\} B, \tag{22.15}$$

where T_n is the noise temperature of the receiver whose bandwidth is B. We see that increasing $g(\theta, \phi)$ by using a more directive aerial always improves the sensitivity. An appreciable increase in $g(\theta, \phi)$ above the maximum value $\frac{3}{2}$ for a dipole is, however, only possible if the aerial dimensions are much larger than a wavelength. This is only feasible if the wavelength is short, when it is also very desirable because of the $1/\lambda^2$ dependence of $S_{\min}$ on wavelength. The importance of the efficiency η also depends on the wavelength. In the microwave region the effective sky temperature is low, a few degrees Kelvin, but below 300 MHz ($\lambda = 1$ m) it increases with wavelength as the effects of ionospheric noise are felt. At 10 MHz it may exceed 500 T_0, and in this case, unless the amplifier noise temperature is quite excessively high, low values of η have little effect. Fortunately at short wavelengths it is much easier to make highly directive and efficient aerials than at long wavelengths, where they are not usually needed.

When the dimensions of the aerial are much less than λ its radiation resistance is negligible and its properties are mainly determined by the open-circuit voltage at its terminals and the properties of the matching network. We give one example. Many broadcast receivers for the 500–1500 kHz band use a ferrite-rod aerial consisting of a rod of effective permeability $\mu \sim 20$, radius $a \sim 5$ mm, and length $l \sim 15$ cm, which is much less than a wavelength. If the rod is wound with N turns the open-circuit voltage with the rod parallel to the magnetic vector **H** of a wave is

$$V = \omega\mu\mu_0\pi a^2NH,$$

and, if this coil is used as part of a tuned circuit of quality factor Q, the signal voltage across the tuning capacitor is $V' = QV$. We can express this in terms of the incident energy flux as

$$V' \sim 400Q\mu a^2NS^{\frac{1}{2}}/\lambda. \tag{22.16}$$

If, for example, $Q = 50$, $\mu = 20$, $a = 5\times10^{-3}$ m, $N = 50$, and $\lambda = 400$ m, we have $V' \sim S^{\frac{1}{2}}$. Thus an incident flux of 10^{-6} W m^{-2} gives about 1 mV. The r.m.s. equivalent noise in a 10 kHz bandwidth with this circuit, which has a shunt impedance of about 50 kΩ, is about 5 μV; thus the signal to noise ratio would be 46 dB with this particular energy flux, which is about what we expect within 100 km of a transmitter of power 100 kW.

22.3. The Dicke radiometer

When white noise of power spectrum w in a bandwidth B_1 is rectified by a square-law detector the output current has a mean value

$$I = \alpha wB_1,$$

and the mean square fluctuations in a narrow bandwidth B_2 are

$$I_n^2 = 2\alpha^2w^2B_1B_2.$$

The change δw in w that gives a change I equal to I_n is therefore

$$\delta w = w(2B_2/B_1)^{\frac{1}{2}}.$$

With a linear detector we have

$$I = \alpha'(wB_1/2\pi)^{\frac{1}{2}}$$

and $$I_n^2 = \alpha'^2wB_1B_2/4\pi$$

so that again $$\delta w = w(2B_2/B_1)^{\frac{1}{2}} \tag{22.17}$$

Thus with either detector we can measure w to a fractional accuracy

$$\delta w/w = (2B_2/B_1)^{\frac{1}{2}}.$$

Consider then an amplifier of bandwidth B_1, noise temperature T_n and power gain A connected to a source at T_s and followed by a detector and a low-frequency amplifier of bandwidth B_2, so that $w = Ak(T_s+T_n)$. If we know A, T_n, and B_1 and these quantities are constant, we can measure changes in T_s of the order of $(T_s+T_n)(2B_2/B_1)^{\frac{1}{2}}$. For example, if $T_s+T_n = 1000$ K, $B_1 = 10^8$ Hz, and $B_2 = \frac{1}{2}$ Hz, the accuracy is 0·1 K.

The need to know A, T_n, and B is eliminated in a system due to Dicke (1946), which is shown schematically in Fig. 22.2. The input is switched cyclically (by S_1) from the source at the unknown temperature T_s to an equal source at a known temperature T_0, and the output is switched in synchronism to either side of an integrating voltmeter V. The bandwidth of this part of the system is $B_2 = 1/4R_1C_1$. If, for simplicity, we assume that the detector is

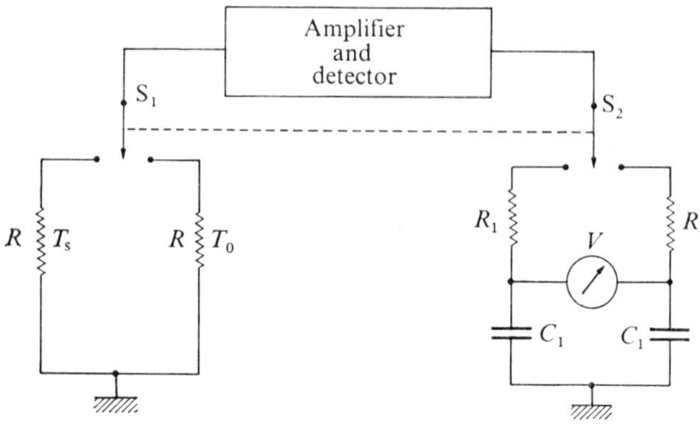

FIG. 22.2. The radiometer circuit

square-law, the integrated voltage developed across C_1, which is connected half the time, is

$$V_1 = \tfrac{1}{2}\alpha AB_1(T_s+T_n)$$

and across C_2 is

$$V_2 = \tfrac{1}{2}\alpha AB_1(T_0+T_n).$$

The mean square noise voltages are

$$V_{1n}^2 = \tfrac{1}{2}\alpha^2A^2(T_s+T_n)^2 2B_1B_2$$

and

$$V_{2n}^2 = \tfrac{1}{2}\alpha^2A^2(T_0+T_n)^2 2B_1B_2.$$

The r.m.s. noise across the voltmeter is therefore

$$V_n = (V_{1n}^2+V_{2n}^2)^{\frac{1}{2}} = \alpha A(B_1B_2)^{\frac{1}{2}}\{(T_0+T_n)^2+(T_s+T_n)^2\}^{\frac{1}{2}},$$

and the least detectable difference between T_s and T_0 is

$$T_s-T_0 = 2\left(\frac{B_2}{B_1}\right)^{\frac{1}{2}}\{(T_0+T_n)^2+(T_s+T_n)^2\}^{\frac{1}{2}} \approx 2\left(\frac{2B_2}{B_1}\right)^{\frac{1}{2}}(T_s+T_n). \quad (22.18)$$

Thus if T_0 is set to give zero output this is the accuracy with which T_s can be measured.

The Dicke radiometer can be used in a number of ways, e.g. as a noise thermometer, but its most important applications are in radio-astronomy.

22.4. Transistor specifications

The noise performance of an f.e.t. at moderate to high frequencies depends on three parameters, the gate leakage current I_g, the mutual conductance g_m, and the input capacitance C or the gain–bandwidth product $g_m/2\pi C$. These parameters are almost always specified in the manufacturers' data sheets, and so there is no difficulty in choosing a particular device or in proceeding with the design of low-noise equipment.

At low frequencies the equivalent noise resistance is increased from $2/3g_m$ to

$$r_n = \frac{2}{3g_m}\left(1+\frac{f_0}{f}\right), \tag{22.19}$$

and the equivalent r.m.s. input noise voltage e_n associated with a bandwidth of 1 Hz is given by

$$e_n^2 = \frac{8kT}{3g_m}\left(1+\frac{f_0}{f}\right). \tag{22.20}$$

In some cases a graph of e_n against f is provided, and the frequency at which e_n is double the high-frequency value gives $f_0/3$. In other cases, e_n is given at a single low frequency, and this can also be used to obtain f_0. If no data is given about the low-frequency noise this usually means that the device is not suitable for use at low frequencies.

The noise performance of bipolar transistors depends on the mutual conductance g_m, the d.c. current gain $\beta_{dc} = h_{FE}$, the gain–bandwidth product f_T, the series base resistance R_b, and the corner frequency f_0 in the expression for the base-current noise

$$w_i = 2eI_b\left(1+\frac{f_0}{f}\right). \tag{22.21}$$

The mutual conductance $g_m = eI_c/kT$ is under the user's control and depends on the bias conditions. Usually β_{dc} and f_T are presented as functions of I_c, thus we need only consider R_b and f_0.

Only very rarely is the base resistance R_b presented in data sheets and, as it exerts an important influence on both the optimum noise temperature and the optimum source resistance, we have to deduce it from other data. This can be done in a number of ways.

At moderate frequencies the optimized noise figure is

$$F_0 = \beta_{dc}^{-\frac{1}{2}}(1+2g_m R_b)^{\frac{1}{2}} = \beta_{dc}^{-\frac{1}{2}}\left(1+2\frac{eI_c R_b}{kT}\right)^{\frac{1}{2}}, \tag{22.22}$$

and if F_0 is given as a function of I_c this can be used in conjunction with the value of β_{dc} to deduce R_b as

$$R_b = \frac{kT}{eI_c}\{(F-1)^2\beta_{dc}-1\}. \tag{22.23}$$

The equivalent r.m.s. input noise voltage, e_n per root-herz, satisfies

$$e_n^2 = \frac{2kT}{g_m} + 4kTR_b\left(1+\frac{g_m R_b}{2\beta_{dc}}\right). \tag{22.24}$$

This can be used to obtain R_b, especially as the minimum value of e_n^2 as a function of I_c is

$$e_{n0}^2 = 4kTR_b\{1+(\tfrac{1}{2}\beta_{dc})^{\frac{1}{2}}\} \sim 4kTR_b \tag{22.25}$$

and occurs when

$$I_c = kT\beta_{dc}^{\frac{1}{2}}/eR_b. \tag{22.26}$$

If data about the low-frequency performance is given it may be in the form of a graph of the equivalent r.m.s. base-current noise i_n per root-herz as a function of I_c and frequency. Since

$$i_n^2 = \frac{2eI_c}{\beta_{dc}}\left(1+\frac{f_0}{f}\right) \tag{22.27}$$

this will yield f_0. In other cases the excess contribution to the noise figure at low frequencies may be presented as $F' = \phi/f$, where ϕ is given as a function of I_c and the source resistance. Now

$$F' = \frac{g_m f_0}{2\beta_{dc} f}\frac{(R_s+R_b)^2}{R_s},$$

and so

$$\phi = \frac{g_m f_0}{2\beta_{dc}}\frac{(R_s+R_b)^2}{R_s}, \tag{22.28}$$

and we see that for all frequencies and all collector currents ϕ is a minimum for $R_s = R_b$. This often gives the most reliable estimate of R_b. Clearly it can also be used to obtain f_0.

In some cases enough data is presented to allow us to obtain more than one independent estimate of R_b, this gives some idea of the accuracy of the various estimates. The accuracy appears to be of the order of ± 30 per cent but fortunately this is sufficient for most purposes. We usually only need to know the order of magnitude of R_b. As a rough general rule small-signal transistors with f_T less than 500 MHz tend to have values of R_b around 200 Ω, while higher-frequency transistors have somewhat lower values of R_b.

22.5. Paramagnetic resonance spectrometers

The phenomena of paramagnetic resonance and nuclear resonance (to be discussed in the next section) result in the appearance of a peak in the complex magnetic susceptibility of a material at a frequency which is determined partly by the value of a steady applied magnetic field and partly by the internal structure of the medium. If we express the susceptibility as $\chi' - j\chi''$ in S.I. units (in which case χ' and χ'' are 4π times as large as in c.g.s. units) and the material is placed within a resonant circuit, the change in the quality factor

Q of the circuit is given by

$$\delta(1/Q) = +\eta\chi'', \tag{22.29}$$

where η is a filling factor. It is the ratio of $\int H^2\,\mathrm{d}V$ over the sample to the same integral over the entire circuit. Paramagnetic resonance is usually studied at microwave frequencies and the tuned circuit is a resonant cavity. Nuclear resonance, on the other hand, is usually studied at frequencies below about 400 MHz, and the circuit consists of a coil and a capacitor. The techniques are rather different and so we shall consider them separately.

An incident wave of r.m.s. amplitude $V_\mathrm{i} = (ZP_\mathrm{i})^{\frac{1}{2}}$ in a line of impedance Z coupled to a cavity of quality factor Q and coupling factor Q_e results in a reflected wave of amplitude

$$V_\mathrm{r} = V_\mathrm{i}\frac{1/Q-1/Q_\mathrm{e}}{1/Q+1/Q_\mathrm{e}}; \tag{22.30}$$

the voltage standing wave in the line is

$$r = Q_\mathrm{e}/Q. \tag{22.31}$$

The change in V_r due to a change $\delta(1/Q) = \eta\chi''$ occurring as the result of resonance absorption is

$$\delta V_\mathrm{r} = \frac{4r}{(1+r)^2}\cdot\tfrac{1}{2}QV_\mathrm{i}\,\delta(1/Q), \tag{22.32}$$

and this constitutes the signal. This signal voltage has to be compared with the r.m.s. equivalent noise voltage associated with a receiver of noise figure F and bandwidth B *matched* to the line; this is

$$V_\mathrm{n} = \tfrac{1}{2}(4ZFkT_0B)^{\frac{1}{2}}. \tag{22.33}$$

The minimum detectable signal is therefore

$$\delta V_\mathrm{r} = (ZFkT_0B)^{\frac{1}{2}}, \tag{22.34}$$

this result does not depend on whether δV_r is the sole reflected amplitude or is only a small change in a large amplitude. Therefore it does not depend on the cavity match or the standing wave ratio r, and we can choose Q_e and r to optimize δV_r. This requires $Q_\mathrm{e} \sim Q$, so that $r \sim 1$. As we shall see, we do not want r to be exactly unity but only near enough unity so that

$$\delta V_\mathrm{r} \sim \tfrac{1}{2}QV_\mathrm{i}\delta(1/Q). \tag{22.35}$$

The least detectable change is now given by

$$\eta\chi'' = \delta(1/Q) = 2(FkT_0B/P_\mathrm{i}Q^2)^{\frac{1}{2}}. \tag{22.36}$$

If P_i and Q are fixed we have only to consider how F and B vary with the reflected power and the structure of the detection circuit.

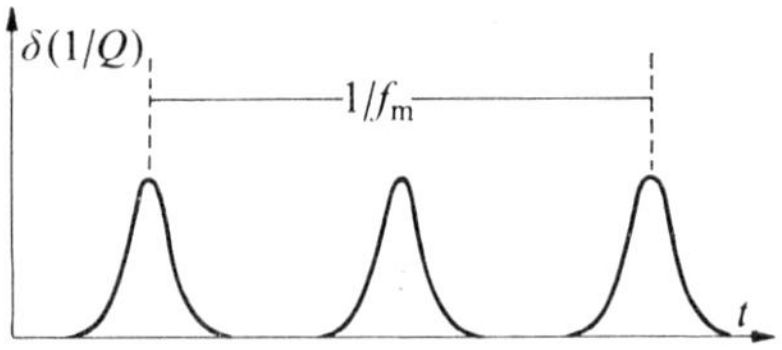

FIG. 22.3. Resonance absorption signal when the field or frequency modulation exceeds the linewidth. The modulation frequency is f_m.

In most spectrometers the absorption signal $\delta(1/Q)$ is modulated by varying the applied magnetic field at a low frequency f_m about the resonance value. If the field sweep appreciably exceeds the width of the resonance line the signal has the form shown in Fig. 22.3 and its repetition frequency is $2f_m$. The receiver bandwidth would have to exceed about $10f_m$ to reproduce this signal. If, on the other hand, the field sweep is less than about half the linewidth and the mean field is set on the side of the line, the signal will be essentially sinusoidal at f_m. Therefore the system could incorporate a final phase-sensitive detector, using the drive to the field modulation as a reference signal, and the bandwidth B can be made as narrow as the stability of the spectrometer permits. This system is shown in outline in Fig. 22.4 The microwave bridge is incorporated so that, whatever the reflection from the cavity, the power delivered to the receiver can be chosen to optimize its performance.

Fig. 22.5 shows a microwave superhet used as the receiver; its noise figure F will depend on the characteristics of the mixer crystal, the intermediate frequency, and the local-oscillator power, but it will not depend on the reflected signal power P_r as long as this does not overload the i.f. amplifier. This must have considerable gain (about 60 dB) to make noise from the second detector negligible, and as a result P_r must be less than about 10^{-8} W. Under these conditions, using a balanced mixer, a noise figure of less than 10 dB can be achieved at 10 GHz.

The bandwidth of the system following the second detector is B_m, the meter bandwidth, provided that the low-frequency amplifier cuts off at less than $3f_m$. However, the effective bandwidth of the whole system is B_m only if the

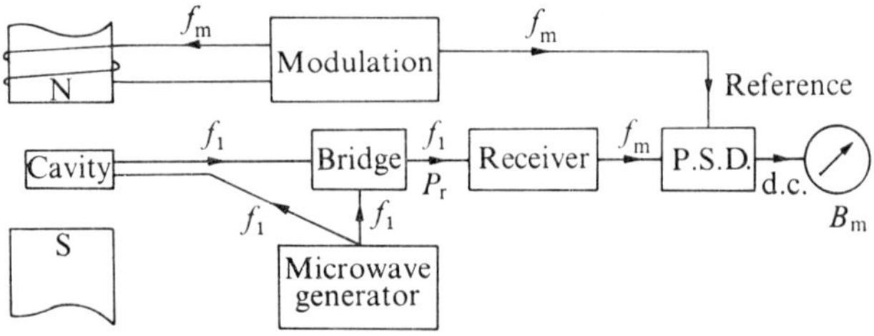

FIG. 22.4. Paramagnetic resonance spectrometer.

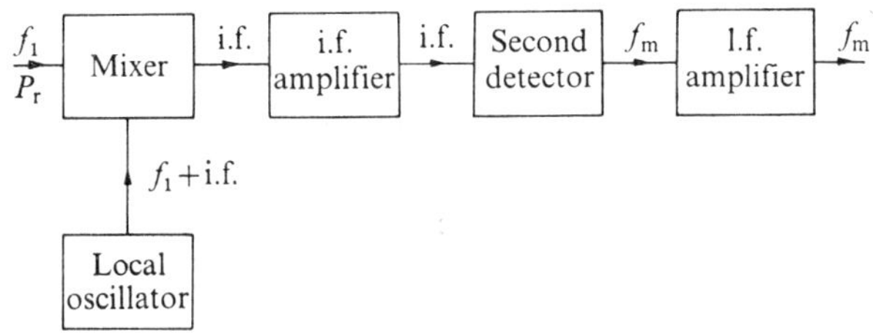

FIG. 22.5. Superheterodyne receiver.

carrier level at the second detector is much larger than the noise. If this is not the case it will be $(2B_iB_m)^{\frac{1}{2}}$, where B_i is the i.f. bandwidth. Since B_m is likely to be less than 1 Hz and B_i greater than 1 MHz it is essential to provide enough carrier amplitude, and this is the function of the microwave bridge. If in (22.36) we put $F = 10$, $B = 0{\cdot}1$ Hz, $P_i = 10^{-2}$ W, and $Q = 5000$ we obtain a limiting sensitivity for the r.m.s. modulation $\delta(1/Q) \sim 2{\cdot}5\times10^{-13}$. The peak to peak modulation is $\eta\chi''$, and so the minimum detectable value of $\eta\chi''$ is about 10^{-12}.

When such extreme sensitivity is not required the apparatus is often simplified by replacing the superhet by a simple crystal diode detector and a low-frequency amplifier, as shown in Fig. 22.6. The bandwidth B is now the bandwidth of the low-frequency amplifier or, if this is followed by a phase-sensitive detector, the bandwidth of the meter. In eqn (20.45) we gave an expression for the effective noise figure of a crystal detector in terms of the incident power P_r, the noise figure F_0 of the low-frequency amplifier, and the crystal parameters μ, η, and γ. This was

$$F = \left(\frac{1+\mu P_r}{2+\mu P_r}\right)^2\left\{\frac{2F_0}{\eta P_r}(1+\mu P_r)+\frac{\gamma P_r}{f_m}\right\}. \tag{22.37}$$

With typical values $\gamma = 10^{12}$ (W s)$^{-1}$, $\eta = 10^3$ W^{-1}, $\mu = 2\times10^3$ W^{-1}, $f_m = 50$ Hz, and $P_r = 10^{-3}$ W (corresponding to $P_i = 10^{-2}$ W and a standing wave ratio $r = 2$), we obtain $F \sim 10^7$, giving a sensitivity, to the small imaginary component χ'' of the susceptibility, some thousand times worse than a superhet. If, however, P_r is adjusted using the microwave bridge to

$$P_r = \left(\frac{2F_0 f_m}{\eta\gamma}\right)^{\frac{1}{2}}, \tag{22.38a}$$

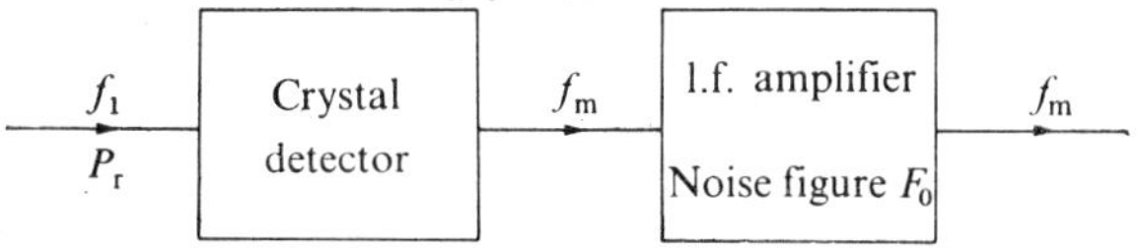

FIG. 22.6. Straight detection system.

we have

$$F = \frac{\mu F_0}{2\eta} + \left(\frac{\gamma F_0}{2\eta f_m}\right)^{\frac{1}{2}}. \tag{22.38b}$$

With the same crystal parameters and $f_m = 50$ Hz this yields $F \sim 3000$. If f_m is increased to 100 kHz the optimum power is increased from $\frac{1}{3}$ μW to 14 μW and F is decreased to 70. The sensitivity is then only about 3 times worse than a superhet.

Thus the best sensitivity is obtained with a superhet, but tolerable sensitivity can also be obtained using a high modulation frequency. In addition, whereas in a superheterodyne system the microwave bridge has to be carefully balanced to reduce the input carrier power to 10^{-8} W, in the straight detection system the required balance to yield 10^{-5} W or so is much less critical.

The minimum detectable value of χ'' depends on the input power and Q in the combination $P_i^{-\frac{1}{2}}Q^{-1}$, and in some cases P_i will only be limited by the availability of a strong c.w. oscillator. In others, the resonance will saturate at some limiting value of the r.f. magnetic field and therefore of the energy density in the cavity. Since the energy density is proportional to QP the sensitivity now varies as $Q^{-\frac{1}{2}}$ rather than as Q^{-1}. In some cases, especially with strong absorption lines, we may find it convenient to dispense with a cavity and use instead a simple transmission system.

Consider an otherwise lossless transmission line filled with a medium for which $\chi = \chi' - j\chi''$ and both χ' and χ'' are small. The power absorbed per unit volume in the medium is $\frac{1}{2}\omega\mu_0\chi''H^2$, where H is the r.f. magnetic field. Thus, if A is the effective cross-section of the system, the rate of decrease of the power with distance is

$$\frac{dP}{dx} = -\tfrac{1}{2}\omega\mu_0\chi''H^2A.$$

But the power itself is given by $\frac{1}{2}\mu_0 H^2 A v_g$ where v_g is the group velocity, and so

$$\frac{dP}{dx} = -\frac{\omega}{v_g}\chi''P.$$

Thus in a distance x, the amplitude is reduced by a factor $\exp(-\omega\chi''x/2v_g)$. The signal voltage, for an input voltage V_i is therefore

$$\delta V \sim \frac{\omega\chi''x}{2v_g}V_i = \frac{\pi x}{\lambda_g}\chi''V_i = \pi n\chi''V_i, \tag{22.39}$$

where λ_g is the wavelength in the transmission circuit and n is the length of the line in terms of wavelengths. If we compare this with (22.32) with $r = 1$ we see that the line is equivalent to a cavity with $\eta Q = 2\pi n$. A helix makes a suitable circuit at wavelengths longer than 2 cm, and it is possible to achieve $n \geqslant 1$ with a helix 1 cm long and 1 mm diameter containing roughly 10 mm^3 of material. Since at a wavelength of 3 cm the cavity would have a volume of about 4 cm³, this amount of material would give $\eta = 2{\cdot}5\times10^{-3}$ and thus we should require $Q \sim 2500$ before the cavity gave better results. In practice, loss in the helix cannot be neglected entirely but nevertheless for many purposes it makes an adequate circuit.

Eqn (22.39), in conjunction with (22.33), gives for the minimum detectable value of χ'' in a transmission system n wavelengths long

$$\chi''_{\text{min}} = \tan\delta = \frac{1}{\pi n}\left(\frac{FkT_0B}{P_{\text{i}}}\right)^{\frac{1}{2}}, \tag{22.40}$$

where δ is the loss angle. In this form (22.40) can also be applied to systems exhibiting dielectric loss as a result of electric, rather than magnetic dipole transitions.

22.6. Nuclear magnetic resonance

Usually nuclear magnetic and quadrupole resonances are studied at frequencies well below 500 MHz, where conventional circuit techniques are applicable and the effects of resonant absorption can be observed as the change in the Q of a tuned circuit when the absorbing medium is placed within the coil.

The power absorbed per unit volume in the medium is $\frac{1}{2}\omega\mu_0\chi''H_1^2$, where H_1 is the peak magnetic field due to the coil. The stored energy per unit volume in the coil is $\frac{1}{2}\mu_0H_1^2$, and so if η, the filling factor is the ratio of specimen volume to coil volume the change in $1/Q$ is

$$\delta(1/Q) = \eta\chi''. \tag{22.41}$$

The corresponding change in the quality factor itself is given by

$$\delta Q/Q = -Q\delta(1/Q),$$

and so the fractional change in the shunt impedance is

$$\delta Z/Z = \delta Q/Q = -Q\eta\chi''. \tag{22.42}$$

If the tuned circuit is driven at resonance from a constant current source this is also the fractional change in the r.f. voltage V_1 across the coil. Thus the signal voltage is

$$\delta V_1 = -Q\eta\chi''V_1. \tag{22.43}$$

The mean square noise voltage across the coil, if it is used in conjunction with an amplifier of noise figure F, is $4kTZF\delta f$, and we can express this in terms of the resonant frequency and the shunt capacitance C in parallel with the coil as

$$V_{\text{n}}^2 = \frac{4FkTQ\,\delta f}{\omega C}. \tag{22.44}$$

The minimum detectable value of $\eta\chi''$ is therefore

$$\eta\chi'' = \left(\frac{4FkT\,\delta f}{Q\omega CV_1^2}\right)^{\frac{1}{2}}, \tag{22.45}$$

where V_1 is the r.m.s. voltage. The r.f. field H_1 in the coil is related to V_1 by

$$CV_1^2 = \tfrac{1}{2}\mu_0 H_1^2 U, \tag{22.46}$$

where U is the coil volume. If saturation effects are significant there will be an upper limit to H_1 for we have

$$\chi'' = \frac{\chi_0''}{1+(H_1/H_r)^2}, \tag{22.47}$$

where χ_0'' is the value of the imaginary susceptibility in zero r.f. field and H_r is fixed by the sample properties. This gives an upper limit to V_1. The allowable values of V_1 vary very considerably, from less than 10^{-4} V to more than 10 V. The optimum value of V_1 is that which makes $H_1 = H_r$, and then we have

$$\eta\chi_0'' = \left(\frac{32FkT\,\delta f}{Q\omega\mu_0 H_r^2 U}\right)^{\frac{1}{2}}. \tag{22.48}$$

We see that large values of η, U, Q, and ω conduce to high sensitivity. As a typical example, for proton magnetic resonance in water in a steady field of 1 T (10^4 G) we might have $H_r \sim 1$ A m^{-1} (0·01 G), $U = 10^{-6}$ m^3 (1 cm^3), $Q = 100$, $F = 2$, $\delta f = 1$ kHz, and $\omega = 2{\cdot}7\times10^8$ s^{-1}, giving $\eta\chi_0'' \sim 10^{-6}$. Since under these circumstances χ_0'' is at least 10^{-3} the signal to noise ratio would be excellent.

When saturation effects are not significant, which, for example, will often be the case with broad lines and in nuclear quadrupole resonance, the value of V_1 will be limited not by saturation but by overloading in the electronic system. In these circumstances (22.45) gives a more useful figure of merit than (22.48).

By making the resonant circuit the tank circuit of an oscillator, it is possible to derive a number of practical advantages: ease of tuning, freedom from microphonic noise, and the possibility of frequency modulating the system. Circuits of this type are known as marginal oscillators. The original circuits, of which perhaps the best known is the Pound, Knight, and Watkins spectrometer, (Pound 1952) used oscillators working in the van der Pol regime. The behaviour of a van der Pol oscillator is, however, rather unpredictable, since the amplitude of oscillation depends critically on the curvature of the characteristics of the active devices in the regenerative feedback loop. Oscillators incorporating a limiter (Robinson 1959) give more stable and predictable performance and equivalent or better sensitivity, especially at either very low amplitudes or very high amplitudes.

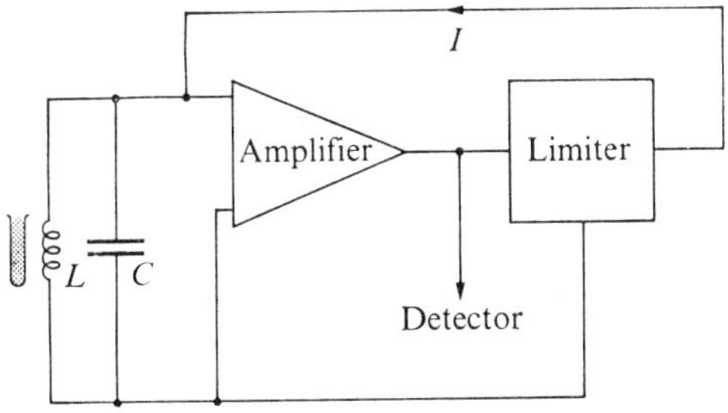

FIG. 22.7. Limited oscillator.

The principle of operation is shown in Fig. 22.7. The output current from the limiter of r.m.s. amplitude I generates a voltage $V_1 = IZ$ across the tank circuit of shunt impedance Z at resonance. The signal due to absorption is therefore again

$$\delta V_1 = -V_1 Q \eta \chi'', \tag{22.49}$$

and, as we saw in Chapter 18, the mean square equivalent noise voltage at this point is

$$V_n^2 = 4FkTZ\,\delta f, \tag{22.50}$$

where F is the amplifier noise figure. Thus again the sensitivity is given by (22.45) and is the same as that obtained using the same amplifier and detector with a separate oscillator drive. Experimental measurements of both the sensitivity of the system as a spectrometer and of the response to injected calibrated white noise confirm this result.

At the frequencies normally encountered it is possible to combine the functions of the limiter and the detector in one circuit, as shown in Fig. 22.8. The properties of this circuit as a detector have been discussed in Chapter 20 and it is clear that for large enough input drives the output current consists of square waves of amplitude equal to the quiescent current in each transistor.

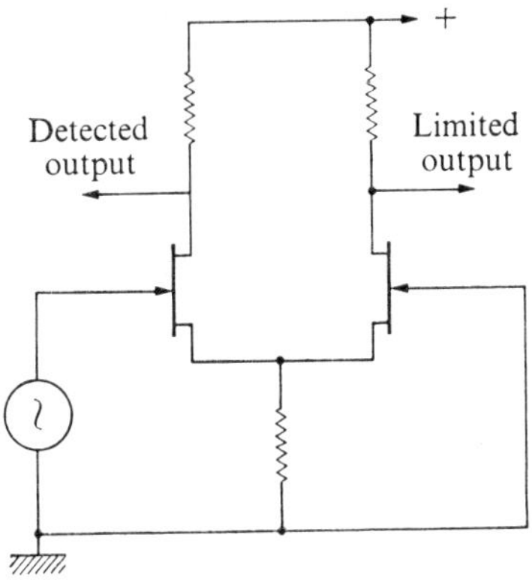

FIG. 22.8. Combined limiter and detector.

Provided that the r.f. level across the tank circuit is low, less than about $\frac{1}{2}$V, so that r.f. amplification can be provided without overloading between the tank coil and the limiter, the detection efficiency is excellent and the sensitivity is almost entirely determined by the noise figure of the r.f. amplifier. If this is an f.e.t. input stage the noise figure need not be appreciably greater than unity. If the required r.f. level is higher, amplification becomes more difficult, and the simple circuit shown in Fig. 22.9 has to be used. Optimum

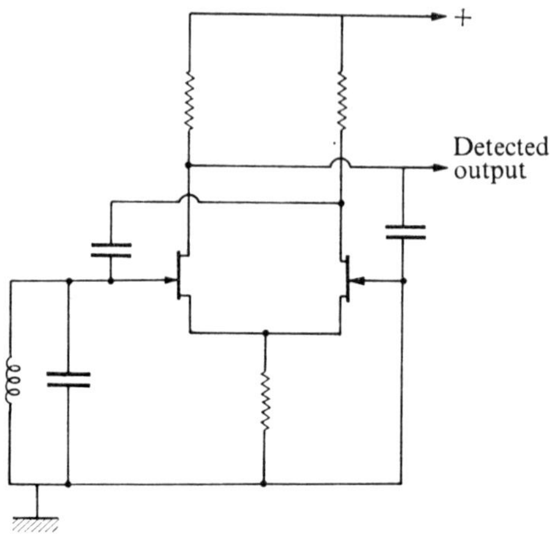

FIG. 22.9. Simple n.m.r. oscillator.

performance as a detector requires relatively low transistor currents, and this conflicts with both the need to obtain sufficient regeneration and the need to provide a large enough limited output current to drive the tank circuit to a high amplitude of oscillation. The conflict is least if the coil Q is high and the tank circuit has a high shunt impedance. It is relatively easy to achieve effective noise figures less than 20 and, especially with large volume high-Q coils, considerably lower figures can be achieved at not too high frequencies. For most purposes this is adequate. It is, for example, easy to see pure nitrogen quadrupole resonance from ^{14}N in hexamethylene tetramine with signals presented on an oscilloscope. Above about 30 MHz, as the tank-circuit impedance falls the sensitivity deteriorates, but nevertheless the circuit still retains a sensitivity of the order of $\delta(1/Q) \sim 10^{-7}$, with a 1 kHz post-detection bandwidth, at 500 MHz. This is, for example, adequate for detecting proton nuclear resonance in very high but highly inhomogeneous fields.

22.7. Practical considerations

The ultimate measure of the quality of an electronic system responding to weak signals is the signal to noise ratio in the final output. This is determined by the effective noise bandwidth of the system and the extent to which noise is present in the input and noise is generated in the system. In an ideal system the effective noise bandwidth is no greater than is necessary to acquire, transmit, and amplify the signal, and no noise is generated within the system. The effective bandwidth is determined by the structure of the system, i.e. the arrangement of amplifiers, mixers, detectors, and filters within the system and the way the output is finally presented or processed. The noise generated within the system is determined almost entirely by the internal sources of noise in the early stages of the system. The degree to which the effective noise bandwidth can be minimized depends on the nature of the signal and the nature of the required output, and we may distinguish between two different cases with respect to both the input and the output. In some cases, e.g. broadcast radio reception, the nature of the incoming signal is fixed by circumstances beyond the receiver designer's control, and in other cases some, or all, aspects of the form of the signal can be controlled. Thus, in a nuclear resonance spectrometer the nature of the applied field (or frequency) modulation, and therefore the signal waveform, can be chosen by the designer, and in space communication the designer can aim to produce a signal which is easy to process. The two different cases applying to the output are concerned with whether, as in telephony and broadcast radio, the output is required instantaneously and so must be processed in real time, or whether some delay is acceptable so that the output can be analysed or processed at leisure. Clearly, the best performance can be achieved when the output is not required instantaneously and the signal can be encoded to match an optimized processing strategy. The form of this strategy is only limited by the ingenuity of the designer. If, however, the signal is required instantaneously and the processing has to occur in real time, the strategy will be limited both by fundamental considerations, e.g. Kramers–Kronig relations and causality, and by the feasibility of the circuits required. The choice of an over-all processing strategy for a system is determined to some extent by the availability of devices for each stage of the system, but these considerations almost always lead to the conclusion that the first stage of a system should be a low-noise, relatively broad-band amplifier accepting all the frequencies associated with the signal. Any more complicated signal processing, e.g. filtering, mixing, or detecting, can be done at a level where noise or attenuation due to these stages can be ignored. The main practical problems associated specifically with low-noise equipment are connected with the design and construction of this pre-amplifier stage, and we shall concentrate our attention on this topic alone.

In the rest of this book we have been concerned almost exclusively with obtaining theoretical expressions for the limiting sensitivity of circuits and devices. These considerations will lead us to the correct choice of a device for a particular application and a circuit design. We now consider what practical precautions have to be taken to achieve the expected performance.

The choice of the active device for the input stage will depend both on the frequency of operation and the impedance of the source. If, for example, the source impedance is 500 Ω, it may be possible to transform it to match a particular device, but equally it may be possible to choose a device for which 500 Ω is ideal. Since all transformers are, to some extent, lossy and generate Johnson noise, in principle at least, the second choice would be better. In practice, it may also be better for another reason, for every additional component in the input is potentially both a possible source of faults and a possible source of interference picked up from outside the system. As a general rule the input circuit design should be as simple as possible and, in particular, should not be liberally adorned with additional frequency-compensating, neutralizing, or temperature-compensating elements. Each additional component is a potential source of trouble and additional noise.

The reasons why a practical circuit may not achieve the expected performance are numerous, but we can list some of the more important ones:

1. There is a fault in the design, e.g. noise generated in a bias resistor or a feedback resistor has been incorrectly ignored.
2. The components in the practical circuit do not correspond to those in the design. It is easy to connect the leads of a transistor incorrectly or to mistake a 47 kΩ resistor for a 470 Ω resistor. A check of the d.c. bias voltages at appropriate points in the circuit will usually identify this type of mistake.
3. The power supply is noisy, and this noise is getting into the first stage via a bias network.
4. Blocking or bypass capacitors are leaky and generating flicker noise.
5. Extraneous noise or signals are being picked up from radio transmitters, other equipment, or the mains.
6. Parasitic oscillations are present at frequencies outside the normal range of operation and the range of the test equipment
7. There are poor contacts, either in the wiring or within components. These may either generate flicker noise or affect the performance directly.
8. The system is mechanically unstable, and sound and vibration are producing microphonic noise. In a low-frequency amplifier this will produce direct effect on the output, in a radio-frequency amplifier it may produce spurious modulation of a carrier.
9. There are faulty components.

The remedies for 1, 2, and 3 are obvious, and 4 can be eliminated by a careful choice of components. The identification of faulty components (9) is often difficult, for small effects e.g. excess collector to base leakage current in a transistor, can have disastrous effects on the noise without appreciably affecting either d.c. conditions or the signal response. Microphonic noise can be eliminated by using a robust mechanical design and keeping all leads short. This, as we shall see, is desirable for other reasons.

In most laboratory locations the ambient electric field strength due to radio and television transmitters and other apparatus, considerably exceeds 1 mV m^{-1}, and if any appreciable voltage is picked up in the input stage of a high-gain amplifier it may either appear in the output or cause overloading of later stages. The resulting non-linear response may then mix the extraneous input with the wanted signal. High-frequency interference can be excluded by careful screening and filtering of the bias supply leads, but the degree of screening required will be much reduced if all the wiring is kept short and the layout is compact. Short leads and the avoidance of all earth loops will also minimize inductive pick-up at low frequencies. In low-noise amplifiers at any frequency it is essential that there should be a well-defined earth plane and that the circuit should be so wired that circulating currents in each stage are confined to loops of the smallest possible dimensions.

Many of the devices used in relatively low-frequency amplifiers have appreciable gain at much higher frequencies. The fact that an amplifier has been designed for use at 1 MHz of itself will not ensure that it cannot oscillate at 1000 MHz. The existence of weak parasitic oscillations at frequencies well outside the expected frequency range is one of the commonest causes of poor noise performance. They usually arise because the odd centimetre of wire in a collector lead or a bypass capacitor lead, though unobjectionable at the design frequency, may represent a high Q inductance at some frequency at which the active device has power gain. There are two remedies for this. First the circuit designer should consider the possible behaviour of his design at all frequencies, secondly the constructor should keep all leads short and avoid earth loops. Perhaps the most familiar example of this problem is the tendency of source-follower, or emitter-follower, stages to operate as Colpitts oscillators if, at any frequency, the input impedance presented to the device is inductive. The presence of parasitic oscillations is usually indicated whenever the performance of an amplifier is altered radically by slight changes in the layout or by touching the wiring with a screwdriver.

Contact resistance can introduce noise, even if the resistance is stable, for it can generate flicker noise. The most usual problem is however with a contact resistance that is itself fluctuating. If, for example, the emitter lead of an input transistor contains a fluctuating contact resistance of the order of 1 mΩ and the bias current is 1 mA this will produce noise equivalent to 1 μV at the input. The effect is even more serious if the resistance occurs within a resonant tuned circuit with an appreciable r.f. voltage across it. In this case, 1 mΩ, which might easily arise from a variable tuning capacitor, in a circuit with 1 V across it can generate several millivolts. This is a particularly serious problem in nuclear magnetic resonance apparatus. The main remedy is to use components with firmly attached (welded or soldered) leads and to ensure that all joints in the wiring are sound. A common source of contact-resistance noise is the contact between a solder tag and an oxidized aluminium chassis.

It will be apparent from these remarks that an amplifier is most likely to achieve its design performance if its mechanical construction is robust, it is well screened, all leads are kept short, there is a well-defined earth plane, and the passive components are chosen with care. These considerations do not usually favour printed-circuit wiring techniques of the conventional variety.

The accuracy of noise calculations is such that a properly designed and constructed amplifier should yield approximately the expected performance.

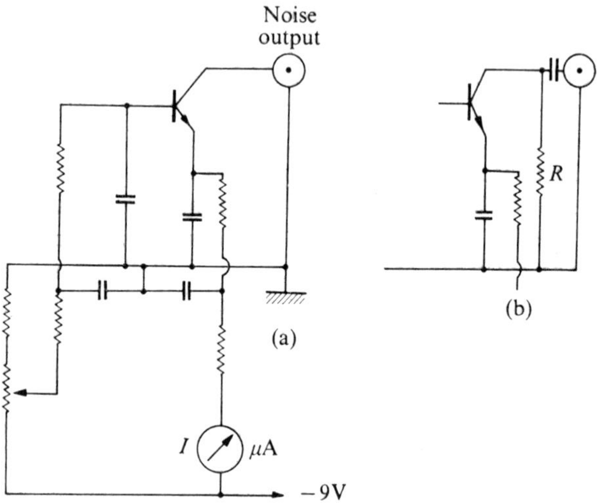

FIG. 22.10(a) and (b). Use of a transistor as a noise source.

If the sensitivity or noise figure is only half the expected value this almost certainly indicates a fault. In the author's experience practical circuits tend either to yield very nearly the expected performance or very much worse performance. Intermediate cases are rare. Thus, although in order to characterize the performance accurately a large range of instruments is needed, it is useful to have a simple method of making a crude check of the noise figure. This is done most easily by injecting a more or less known amount of noise into the input. A useful source is the shot noise in the collector current of a transistor with its base capacitatively shorted to the emitter. The basic circuit is shown in Fig. 22.10(a). The meter is used to read the transistor current which is controlled by the potentiometer. If it were not for the series base resistance R_b the power spectrum of the output noise current would be $2eI$; the presence of the base resistance increases it. Thus if this source is connected to the input of an amplifier with a source impedance R_s also in position and I is the current required to double the output noise, the noise figure cannot be less than $eIR_s/2kT \sim 20IR_s$. The accuracy of this technique is not high, but it is enough to indicate a major departure from the expected performance. Note that if R_s is 10 kΩ and the expected noise figure is near unity the current required is only 5 μA. Fig. 22.10(b) shows a modification which may be necessary if the d.c. transistor current must be kept out of the amplifier input stage. The value of the resistance R should be chosen either to equal the correct source impedance R_s or be made much larger than R_s, if R_s is already in place. It is, of course, important to ensure that the voltage drop in R is not sufficient to remove the collector bias from the transistor.

Transistor amplifiers are not linear devices; the bipolar transistor has an exponential transfer characteristic, and the f.e.t. has a quadratic characteristic. If θ is the fractional modulation of the mean transistor current, the ratio of the quadratic term to the linear term in a power series expansion of the transfer characteristic is $\theta/2$ for the bipolar device and $\theta/4$ for the field-effect device. In a complete electronic system we usually aim to raise the signal and noise level to at least 50 mV at the output of the pre-amplifier, and in many cases this is then connected to the next stage by a low-impedance line. If the line impedance is, say, 50 Ω the required r.m.s. output current from the pre-amplifier is 1 mA, which may be an appreciable fraction of the standing current in the last transistor. It is then possible for noise or unwanted signals from outside the signal bandwidth to become mixed into this bandwidth and produce additional noise in the final output. It is not difficult to avoid this effect, but clearly it must be considered when the initial design calculations are made.

In superheterodyne receivers an amplified, but still weak, signal is essentially switched by the local-oscillator drive to the mixer. The mixer output consists of the weak intermediate frequency output at f_i and a number of other components, but especially a strong component at the local-oscillator frequency f_0. This will be several orders of magnitude larger than the intermediate frequency output and, if it is not filtered out, can easily overload the first intermediate frequency stage. The problem is especially acute in broadcast radio receivers, where f_0 may be as little as twice f_i and the intermediate frequency tuned circuits have a relatively low $Q = f_i/\Delta f_i \sim 50$. If the first i.f. stage is seriously overloaded there will be an increase in the noise, and the signal will be reduced and possibly made unrecognizable.

In many systems the final bandwidth is quite narrow but the bandwidth up to the last stage is much larger. For example, in a system terminating in a phase-sensitive detector the final bandwidth may be only 1 Hz, but the bandwidth immediately prior to the detector may be 10^4 Hz. The signal level at this point must be at least sufficient to give an appreciable output, and the noise may be much larger. Thus, if a threshold signal gives a detector output of one-tenth full scale, the r.m.s. noise level at the input will correspond to a signal amplitude giving 10 times full-scale output. Noise peaks of amplitude several times the r.m.s. value have an appreciable probability of occurring, and so care will have to be taken to ensure that overloading does not occur at this point.

A wide variety of packaged low-noise amplifiers and low-noise integrated circuits is now available, and often these can be used instead of a new design. It is, however, essential to be sure that the amplifier is suitable for the particular application. Not only must the noise performance be good in the frequency range of interest, but also the required source impedance must be correct. The noise performance of many amplifiers is specified in terms of the equivalent r.m.s. input noise voltage per root-herz. This by itself is not an adequate figure of merit unless it is accompanied by information about the equivalent input-noise current or information about the variation of the noise performance with source impedance.

Finally, we note that an electronic system can only amplify and process the signal actually presented to its input terminals. Since cables have an appreciable attenuation at high frequencies and can also pick up interference, it is often desirable to mount the pre-amplifier near or on the signal source. This is facilitated if the pre-amplifier is compact. Since a compact design both encourages and requires careful layout and short wiring it is usually good practice to make the pre-amplifier as small as possible. This should be born in mind especially if use is made of a standardized printed card or similar system. These systems are usually designed to accommodate much more complex circuits, and the space available on a typical card is an invitation to careless layout and wiring.

Appendix A: The current induced by a moving charge

THE formula

$$I = e\mathbf{v} \,.\, \mathbf{E}/\phi_0. \tag{A1}$$

for the current I induced in the leads to two electrodes at a potential difference ϕ_0 producing a field $\mathbf{E}$, when a particle of charge e and velocity $\mathbf{v}$ moves in their vicinity, is used in § 5.3 and § 9.1, and is there derived from an energy-balance equation. This is not entirely satisfactory, as the energy-balance equation that we used omitted any change in the field energy of the system due to the motion of the charge. We now give a more rigorous discussion.

We consider two electrodes, one earthed and the other maintained at a steady potential ϕ_0 by a battery. These electrodes produce a field $\mathbf{E} = -\nabla\phi$. If any dielectric media are present, we assume that they are linear so that $\mathbf{D} = \epsilon\epsilon_0 E$, where ϵ, though a function of position (and also possibly anisotropic), does not depend on $\mathbf{E}$. Charged particles move near the electrodes and constitute a current density $\mathbf{J}$. Then, provided that magnetic effects can be ignored, the equation

$$I\phi_0 = \int \mathbf{E} \,.\, (\mathbf{J}+\dot{\mathbf{D}})\,\mathrm{d}V, \tag{A2}$$

where I is the current drawn from the battery, is both the complete energy-balance equation and a direct consequence of the field equations (Robinson, 1973). Now let ϕ^*, $\mathbf{E}^*$, and $\mathbf{D}^*$ be the potential, field, and displacement due to the same electrodes at the same potentials with the same dielectric media present but with no mobile charge carriers present, so that $\nabla \,.\, \mathbf{D}^* = 0$. We have

$$\begin{aligned}\mathbf{E} \,.\, (\mathbf{J}+\dot{\mathbf{D}}) &= (\mathbf{E}-\mathbf{E}^*) \,.\, (\mathbf{J}+\dot{\mathbf{D}})+\mathbf{E}^* \,.\, \mathbf{J}+\mathbf{E}^* \,.\, \dot{\mathbf{D}} \\ &= (\mathbf{E}-\mathbf{E}^*) \,.\, (\mathbf{J}+\dot{\mathbf{D}})+\mathbf{E}^* \,.\, \mathbf{J}+\mathbf{D}^* \,.\, \dot{\mathbf{E}}.\end{aligned}$$

Because $\nabla \,.\, (\mathbf{J}+\dot{\mathbf{D}}) = 0$ and $\nabla \,.\, \mathbf{D}^* = 0$ the volume integral of the first term on the right becomes the surface integral of $-(\phi-\phi^*)(J_\mathrm{n}+\dot{\mathbf{D}}_\mathrm{n})$, and that of the last term becomes the surface integral of $-D^*_\mathrm{n}\dot{\phi}$. Both these surface integrals are zero, the first because $\phi = \phi^*$ everywhere on the surface, the second because ϕ is independent of time on the surface. We therefore have

$$I\phi_0 = \int \mathbf{E} \,.\, (\mathbf{J}+\dot{\mathbf{D}})\,\mathrm{d}V = \int \mathbf{E}^* \,.\, \mathbf{J}\,\mathrm{d}V. \tag{A3}$$

If the current is due to particles of charge e_i moving at velocities $\mathbf{v}_i$ at positions $\mathbf{r}_i$ we have the general result

$$I\phi_0 = \sum_i e_i\mathbf{v}_i \,.\, \mathbf{E}^*(\mathbf{r}_i). \tag{A4}$$

It gives the induced current

$$I_i = \frac{e_i}{\phi_0} \mathbf{v}_i \cdot \mathbf{E}^*(\mathbf{r}_i), \tag{A5}$$

due to the motion of a single charge, provided that, during this motion, there is no consequent redistribution of the other charges. It would be the correct formula to use to obtain the current due to the transit of a carrier across the depletion layer of a semiconductor junction diode. Thus in § 9.1 the field should be interpreted as $\mathbf{E}^*$ (the field in the depletion layer in the absence of both the unneutralized donor and acceptor ions and the small number of mobile carriers). If the depletion layer is planar and of width d, we have $\mathbf{E}^* = \phi_0/d$ and so $I_i = ev_i/d$. We did not draw attention to this point in § 9.1 partly to avoid breaking the chain of the argument and partly because the use made of the relation in this section does not depend on the distinction between $\mathbf{E}^*$ and $\mathbf{E}$. As long as we only need to know that the transit of a single electron produces a short pulse of total charge e the distinction between $\mathbf{E}^*$ and $\mathbf{E}$ is irrelevant. The use of $\mathbf{E}$ rather than $\mathbf{E}^*$ merely leads to an incorrect shape for the pulse.

Although eqn (A4) for the total current due to all the charges is correct in all circumstances, eqn (A5) for the current due to a single charge is less general. In a conducting medium the remaining charges move in such a way as to maintain charge neutrality as a single charge moves. Thus they tend to reduce the effect of this charge. However, if charge neutrality is maintained, we also have $\dot{\mathbf{D}} = 0$ and so the equation

$$I\phi_0 = \sum_i e_i \mathbf{v}_i \cdot \mathbf{E}(\mathbf{r}_i). \tag{A6}$$

gives the same result as (A4), even though $\mathbf{E}$ may be very different from $\mathbf{E}^*$. Furthermore, since $\dot{\mathbf{D}} = 0$, the field energy is constant and the energy-balance equation for a single charge is exactly

$$I_i\phi_0 = e_i\mathbf{v}_i \cdot \mathbf{E}_i(\mathbf{r}_i). \tag{A7}$$

The current I_i given by (A7) contains both the direct contribution $e_i\mathbf{v}_i \cdot \mathbf{E}^*/\phi_0$ and the compensating current due to the screening effect of the other charges. Thus in § 5.3 the use of the actual electric field $\mathbf{E}$ in the conducting medium is compatible with the rest of the arument which rests on the interpretation of (A7) as the true energy-balance equation.

Appendix B: Noise in *p–n* junction

For simplicity we consider only a diode in which carrier recombination in the depletion layer can be ignored, so that its static characteristic is

$$I = I_0\{\exp(eV/kT)-1\}. \tag{B1}$$

The essential point made by Buckingham and Faulkner (1974) is that the forward flux I_F of carriers across the depletion layer is not equal to $I_0 \exp(eV/kT)$, nor is the reverse flux I_R equal to I_0. Both these fluxes are much larger than I_0, by a factor which is about the ratio of the diffusion length to the mean free path between collisions, and (B1) describes only the small difference between I_F and I_R. Whereas these large currents are solely controlled by the carrier concentrations in the immediate vicinity of the depletion layer, the currents that appear in (B1) are controlled by the rate at which minority carriers, throughout the bulk material, can diffuse towards and away from the depletion layer. Although the fluctuations in the external diode current are still directly related to fluctuations in the rate at which carriers cross the depletion layer, these fluctuations themselves are not simply due to shot noise in I_F and I_R. Because the bulk diffusion process is the bottleneck in the flow of charge carriers, fluctuations in the large currents I_F and I_R lead to charge accumulation, and this alters the field across the depletion layer so as to annul the external effects of these fluctuations. As Buckingham and Faulkner show, this process is so effective that shot noise in I_F and I_R makes almost no contribution to the external diode noise. This arises almost entirely from fluctuations in the rate at which minority carriers diffuse back and forth in the bulk material, and thus control the net carrier flux across the depletion layer.

Within the bulk material majority-carrier motion maintains overall charge neutrality and the minority-carrier motion is entirely due to diffusion. A solution of the time-dependent diffusion equation in the bulk material yields an expression for the mean minority-carrier flux, in terms of the minority-carrier concentrations p_0 and p_w at the two boundaries of the bulk region. Since p_0, at the edge of the depletion layer, responds to the applied bias voltage, this expression in turn yields the junction admittance $Y(\omega) = G(\omega)+jB(\omega)$.

The motion of a single minority carrier between collisions in the bulk region leads to a local alteration in the minority-carrier concentration and its effects diffuse to the boundaries of the bulk region. The total current at the edge of the depletion layer can be expressed as a sum of the effects of all the carrier movements within the bulk region. If the individual carrier movements are treated as independent random events the sum of the squares of their deviations from their mean behaviour yields the noise. We shall not give the details of Buckingham and Faulkner's ingenious calculation but only its most important results.

Their expression for the power spectrum of the noise in a forward-biased junction diode is

$$w = 2eI + 4kT(G(\omega) - G(0)), \tag{B2}$$

and this is identical with equation (9.8a). As long as we regard $G(\omega)$ as an empirical quantity the results of both theories are the same. However, because Buckingham and Faulkner relate $G(\omega)$ to the high-frequency response of the diffusion process, rather than to our somewhat contrived notion of multiple transits across the depletion layer, the connection between $G(\omega)$ and the physical parameters of the junction is different.

Their calculations of the effects of recombination in the depletion layer and of collector-current noise in a transistor are identical with those given in Chapters 9 and 10. Their calculation of the low-frequency base current noise due to recombination in the base is also equivalent to that in Chapter 10. Since (B2) is the same as (9.8a) we might expect their final results to be identical. This is not so. Although, in both treatments the high-frequency increase in the base-current noise is ultimately due to the random time delay between fluctuations in the collector and emitter currents, the nature of the calculation is different. A similar difference occurs in the calculation of the high-frequency effects of recombination events in the base. Their results are most easily compared with those given in Chapter 10, and used in Chapter 13, by considering the leading terms, in an expansion in powers of f/f_α, of the expressions for the base-current noise and its correlation with the collector-current noise. The relevant formulae from Chapter 10 are (10.1d) and (10.1e), and these yield

$$w_b = 2eI_b + 4\left(\frac{f}{f_\alpha}\right)^2 eI_c, \qquad w_{bc} = -2j\left(\frac{f}{f_\alpha}\right) eI_c. \tag{B3}$$

Buckingham and Faulkner's results are

$$w_b = 2eI_b + \frac{2}{3}\left(\frac{f}{f_\alpha}\right)^2 eI_c, w_{bc} = -\frac{2}{3}j\left(\frac{f}{f_\alpha}\right) eI_c. \tag{B4}$$

In both cases the power spectrum of the collector current in $2eI_c$. We see that their calculation leads to less high-frequency noise and that the validity of the low-frequency approximation $w_b = 2eI_b$, $w_{bc} = 0$ is extended to higher frequencies, of the order $(3/\beta)^{\frac{1}{2}} f_\alpha$ rather than $(1/2\beta)^{\frac{1}{2}} f_\alpha$.

In practice, since at frequencies where the difference between (B3) and (B4) is appreciable the noise performance is dominated by the rather inaccessible parameter R_b, the base resistance, the consequences of their results are not, as they remark, as significant as the rather large differences between the numerical factors might suggest. For design purposes a rough and ready correction to the relevant formulae in Chapters 10 and 13 is to replace f_T by $2f_T$.

Apart from its sounder physical interpretation of the noise processes in p–n junctions, Buckingham and Faulkner's analysis has other advantages. It relates the admittance $Y(\omega)$ of the emitter–base junction to accessible transistor parameters and, as they show, gives a simple explantation of the increased noise observed in a saturated transistor whose collector is bottomed.

Now that we attribute the origin of junction noise to diffusion processes in the bulk material, rather than to shot-noise processes in the primary carrier flux across the depletion layer, some of the statements in Chapter 5 need qualification.

References

ADLER, R., HRBEK, G., and WADE, G. (1959). *Proc. Inst. Radio Engrs* **47,** 1713.

BAECHTHOLD, W., WALTER, W., and WOLF, P. (1972) *Electronics Lett.* **8,** 35.

BELLMAN, R. (Ed.) (1961). *Modern mathematical classics: analysis.* Dover, New York.

BUCKINGHAM, M. J. and FAULKNER, E. A. *The Radio and Electronic Engineer* Vol. 44, No. 3 March 1974 pages 125–140. "*The Theory of inherent noise in p–n junction diodes and bipolar transistors*".

BURGESS, R. E. (1941). *Proc. phys. Soc.* **53,** 293.

CAMPBELL, N R. (1909). *Proc. Camb. phil. Soc. Math. Phys. Sci.* **15,** 117, 310.

CHU, L. J. (1951). *Inst. Radio Engrs Conference, June, New Hampshire.*

DICKE, R. H. (1946). *Rev. Sci. Instr.* **17,** 268.

DIRAC, P. A. M. (1958). *Principles of Quantum Mechanics.* Oxford University Press, London.

DRAGONE, C. (1972). *Bell Syst. tech. J.* **51,** 2169.

EINSTEIN, A. (1906). *Annln Phys.* **19,** 289, 371.

FAULKNER, E. A. and HOLMAN, A. (1967). *J. scient. Instrum.* **44,** 391.

FELLER, W. (1957). *An introduction to probability theory and its applications.* Wiley, New York.

FRIIS, H. T. (1944). *Proc. Inst. Radio Engrs* **32,** 419.

GORDON, J. P., WALKER, L. R., and LOUISELL, W. H. (1963). *Phys. Rev.* **130,** 807.

GOLDSTEIN, H. (1950). *Classical Mechanics.* Addison–Wesley.

HAUS, H. A. and ADLER, R. B. (1959). *Circuit theory of linear noisy networks.* Technology Press, New York.

HEFFNER, H. (1962). *Proc. Inst. Radio Engrs* **50,** 1604.

JOHNSON, J. B. (1928). *Phys. Rev.* **32,** 97.

KHINTCHINE, A. I. (1934). *Math. Annln* **109,** 604.

KITTEL, C. (1958). *Elementary statistical physics.* Wiley, New York.

LORENTZ, H. A. (1952). *Theory of electrons.* Dover, New York.

LOUISELL, W. H. (1964). *Radiation and noise in quantum electronics.* McGraw-Hill, New York.

MACDONALD, D. K. C. (1949). *Phil. Mag.* **40,** 561.

MANLEY, J. M. and ROWE, H. E. (1956). *Proc. Inst. Radio Engrs* **44,** 904; also (1959) **47,** 2115.

MUMFORD, W. W. (1949). *Bell Syst. tech. J.* **28,** 608.

NORTH, D. O. and FERRIS, W. R. (1941). *Proc. Inst. Radio Engrs* **29,** 49.

NYQUIST, H. (1928). *Phys. Rev.* **32,** 110.

PENFIELD, P. and RAFUSE, R. P. (1962). *Varactor applications.* M.I.T. Press.

POUND, R. V. (1952). *Progr. Nucl. Phys.* **2,** 21.

RACK, A. J. (1938). *Bell Syst. tech. J.* **17,** 592.
RICE, S. O. (1944). *Bell Syst. tech. J.* **23,** 282; also (1945) **24,** 46.
ROBINSON, F. N. H. (1952). *Phil. Mag.*, Ser 7, **43,** 51.
———. (1958*a*). *Phil. Mag.* **3,** 909.
———. (1958*b*) *J. Electron. Control* **5,** 152.
———. (1959). *J. scient. Instrum.* **36,** 481.
———. (1965). *Proc. R. Soc.* **A286,** 525.
———. (1973). *Macroscopic electromagnetism.* Pergamon Press, Oxford.
ROWE, H. E. (1965). *Signals and noise in communication systems.* Van Nostrand, New York.
SAH, C.-T., and HIELSCHER, F. H. (1966). *Phys. Rev. Lett.* **17,** 956.
SAH, C.-T., NOYCE, R. M., and SHOCKLEY, W. (1957). *Proc. Inst. Radio Engrs* **45,** 1228.
SCHOTTKY, W. (1937). *Wiss. Veröff. Siemens-Werken* **16,** 1.
SHOCKLEY, W. and PIERCE, J. R. (1938). *Proc. Inst. Radio Engrs* **26,** 321.
SPENKE, E. (1937). *Wiss. Veröff. Siemens-Werken* **16,** 19.
SUHL, H. (1957). *Phys. Rev.* **106,** 384.
THOMPSON, B. J., NORTH, D. O., and HARRIS, W. A. (1940). *RCA Rev.* **4,** 269, 441; **5,** 244, 371, 505; **6,** 114.
VAN DER POL, B. (1927). *Phil Mag.* **3,** 65.
VAN DER ZIEL, A. and BECKING, A. G. T. (1958). *Proc. Inst. Radio Engrs* **46,** 589.
———. (1959). In *Noise in electron devices* (Eds. L. D. Smullin and H. A. Haus). Wiley, New York.
———. (1962). *Proc. Inst. Radio Engrs* **50,** 1808.
———. (1963). *Proc. Inst. Radio Engrs* **51,** 461.
WIENER, N. (1930). *Acta math., Stockh.* **55,** 117.

Bibliography

General texts on noise

BENNETT, W. R. (1960). *Electrical noise*. McGraw-Hill, New York.

DAVENPORT, W. B. and ROOT, W. L. (1958). *Random signals and noise*. McGraw-Hill, New York.

LAWSON, J. L. and UHLENBECK, G. E. (1950). *Threshold signals*. McGraw-Hill, New York.

ROWE, H. E. (1965). *Signals and noise in communication systems*. Van Nostrand, New York.

VAN DER ZIEL, A. (1954). *Noise*. Prentice-Hall, New York.

WAX, N. (Ed.). (1954). *Noise and stochastic processes*. Dover, New York (this contains the classic papers Mathematical analysis of random noise by S. O. Rice, reprinted from *Bell Syst. tech. J.*)

Information theory

BRILLOUIN, L. (1956). *Science and information theory*, Academic Press, New York.

PIERCE, J. R. (1961). *Symbols, signals and noise*. Harper, London.

SHANNON, C. E. and WEAVER, W. (1963). *The mathematical theory of communication*. University of Illinois Press.

Electronic circuits

CHERRY, E. M. and HOOPER, D. E. (1968). *Amplifying devices and low-pass amplifier design*. Wiley, New York.

COWLES, L. G. (1966). *Analysis and design of transistor circuits*. Van Nostrand, New York.

FAULKNER, E. A. (1969). *Introduction to the theory of linear systems*. Chapman and Hall, London.

Statistical mechanics

KITTEL, C. (1958). *Elementary statistical physics*. Wiley, New York.

SCHRÖDINGER, E. (1946). *Statistical thermodynamics*. Cambridge University Press.

Quantum mechanics

DIRAC, P. A. M. (1958). *Principles of quantum mechanics*. Oxford Univeristy Press.

KRAMERS, H. A. (1956). *Quantum mechanics*. Dover, New York.

LOUISELL, W. H. (1964). *Radiation and noise in quantum electronics*. McGraw-Hill, New York.

Noise in vacuum tubes and semiconductor devices

SMULLIN, D. and HAUS, H. A. (Ed). (1959). *Noise in electron devices*. Wiley, New York.

Oscillators

HAFNER, E. (1966). The effects of noise in oscillators, *Proc. Inst. Radio Engrs* **54,** 176–98.

Electron-beam tubes

PIERCE, J. R. (1950). *Traveling wave tubes,* Van Nostrand, New York.

ROBINSON, F. N. H. and HAUS, H. A. (1956). Analysis of noise in electron beams, *J. Electron.*, Ser. 1, **4,** 373–84.

Author Index

Subject Index